The Angler's Glossary

By

Michael J. Klimkos

Library of Congress Control Number 2023905356
ISBN 9798386595494

Back cover quotes

Zern, Ed, *How To Tell Fish From Fishermen or a Plague on Both Your Houses*, D. Appleton-Century Company, New York, 1947

McManus, Patrick F., *Never Sniff a Gift Fish*, Holt, Rinehart and Winston, New York, 1979

Waterman, Charles F., *Reel Clicks and Stiff Leaders, Silent Seasons*, Clark City Press, Livingston, MT., 19978

Cooks Run Publications, Carlisle, Pennsylvania
2023

Also by Michael J. Klimkos

A History of Trout Unlimited and the Environmental Movement in Pennsylvania
The Letort: A Limestone Legacy (Editor)
The Fires of Penn's Woods
Waters of the Valley: 50 Years of Cumberland Valley Trout Unlimited
Fly Hook Pocket Reference Guide

To Luciana and Mila
There is always more to learn.

Acknowledgements

There are a multitude of people that helped me put this book together. To all of them I am truly grateful. Julie at The Essential Fly in Selby, North Yorkshire, England helped me with flies that are common in Europe. Thanks to Ed Jaworowski who graciously provided a fly casting photo, and Ron Seidle, Dan Dunmire and Roy Richardson who provided the talent for a few of the photos. Also thanks to Gary Leone for his photo of a trout and Mary Kuss for her Tenkara fly photo. Tim Trexler provided the photo of the Orange Parson which he also tied. Don Kelly proprietor of The Tackle Shack in Wellsboro, Pennsylvania lent his expertise in ice fishing and provided photos. Jaren Cox and Will Fowler at Frogg Toggs provided photos of boots and waders. Steve Fournier and Bruce Eng at Dr. Slick went out of their way to make sure I got the photos I needed in the proper format. Marcos Vergara at Hareline Dubbin provided photos of tying material and Scott McCann at Cortland Line Company supplied photos of lines and leaders. Justin Pittman and Brian Largent and the staff at Precision Fly and Tackle provided their expertise and took a first look at the manuscript and provided input. Josh Barrett at Holly Flies took time from his day to help me find the photographs of flies that I needed. To all of you a special thank you. To the late Ed Zern, Patrick F. McManus and Charles Waterman who originally wrote the pithy quotes that I used on the back cover. Wherever you gentlemen are I hope your flies and martinis are dry, your casting is precise and your tippets are strong. A special thanks to Lyn Langer and Karl Sheaffer who provided suggestions and guidance through this project. To Tom Sholseth, DVM MPVM who reviewed it, provided insight and the foreword, I am extremely grateful for his assistance. Finally, to my wife Susie, without her support I could not have completed this work.

Foreword

Angling is a pastime with a rich and varied history and as such, it comes with a specialized vocabulary of terms and jargon. Whether you are a seasoned angler or just starting out, understanding the language of fishing is one key to success on the water. Fortunately, Michael Klimkos is an accomplished locksmith.

His book provides an entrance to some of the terminology used in angling. From basic terms like "barb and "shank," to more specialized terms like "estuary" and "profundal zone," this book covers some of what you need to know to communicate effectively with other anglers and understand the nuances of angling and aquatic science.

Each term is defined in clear and concise language, with examples and illustrations where appropriate. The book is organized alphabetically for easy reference, and as such is a glossary of commonly used abbreviations and acronyms.

Whether you are a fly fisherman, a baitcaster, or a spin fisherman, this book will be a valuable resource for expanding your knowledge and understanding of the language of angling. I hope that it will serve as a guide and companion to all those who love this wonderful sport.

Tom Sholseth DVM MPVM
Ladysmith, Vancouver Island, British Columbia, Canada
April, 2023

Preface

This is not a book that will provide instructions on how to fish. This book is intended to define terms to help novice and experienced anglers better understand their sport.

I have been involved with Pennsylvania's Rivers Conservation & Fly Fishing Youth Camp since the original planning meeting and the first camp in 1995. As the former director of curriculum I always tried to provide the students with the best references and resources available. When I "retired" from the camp in 2021 I left the camp and my former office in very capable hands. But like an old fire horse when he hears the fire bell, come May 2022 I found myself wondering what I could do to help the camp.

As part of another project I had been compiling a glossary of terms related to fishing hooks and fishing. I dusted the glossary off, added a few more terms and a couple of photos and sent it on to be distributed to the students. Since then I have been adding both scientific and angling terms to the list. I have included terms for all types of angling, not just fly fishing. I hope to broaden the anglers' knowledge in that respect. The scientific terms are common to many who work in the field, but to the average angler they may not be. However, these terms sometimes show up in articles, web posts and blogs and even in conversation with other anglers.

The definitions found here are not the final words in terms found in fishing or the science behind fish and their environment. This work is instead a small collection of words that may be useful to the novice as well as the advanced angler to provide clarification. Certainly not every term is included.

It is hoped that the reader will continue their adventures in angling, be it with a fly rod, spinning rod, cast net or whatever device they find suitable in fresh or saltwater.

MJK
April 2023

A

Aberdeen Hooks – Hooks that are generally long shank, fine or extra fine wire, with round bends. They can bend under stress.

Figure 1 - Aberdeen Hook with Ring Eye

Acid Mine Drainage or **AMD** – Acidic water with a high content of metals and sulfates, caused by the exposure of rocks containing acid forming minerals to the atmosphere and water due to mining. Iron, manganese and aluminum are the most abundant metals though other metals such as cadmium, zinc, copper, selenium, arsenic and lead may be present as well. Recent studies have found that rare earth metals are also present and AMD may prove to be a source for these strategic metals. AMD is called Acid Rock Drainage in the American West.

Figure 2 - Acid Mine Drainage – AMD may be a source for rare earth metals.

Acidity - A measurable quantity of free hydrogen ions in a solution.

Adfluvial - Fish that spawn in tributary streams and then spend their juvenile growth period (1 - 4 years more or less) before migrating to a lake to reach maturity

Adipose Fin – A small fleshy fin found on the back of a fish, posterior to the dorsal fin. It is found on trout and salmon and only a few other fish.

Aerobic – In the presence of oxygen. *See also Anaerobic*

Alevin - A newly spawned salmon or trout still carrying the yolk sac as a food source. Sometimes called a sac fry.

Algae – Mostly single celled aquatic plants that lack stems, roots, leaves and a vascular system found in higher plants. When certain types of algae overwhelm a water body a Harmful Algae Bloom or HAB occurs. The algae can produce toxins that are harmful to humans and other animals. They may occur in small stagnant freshwater ponds, brackish estuaries or saltwater, where they are known as Red Tides.

Alkalinity - The measurable ability of water to buffer acids and maintain a consistent pH

Allopatry - A population or species that is isolated from a similar group or population by some barrier

Amphidromous – Fish that regularly migrate between saltwater and freshwater but not for the purpose of breeding. Gobies are an example.

Amphipod - An order of crustaceans that are shrimp-like in form such as cress bugs and scuds.

Anadromous - Fish that are born in freshwater migrate to the ocean and return to freshwater to spawn, Salmon and steelhead for example.

Anaerobic – Without oxygen. *See also Aerobic.*

Anal Fin – The single fin on the bottom of the fish closest to the tail. *Appendix D.*

Angle – The act of trying to capture fish. A person who does this is known as an Angler.

Anthropogenic - Caused by human activity.

Anti-Reverse – A reel where the spool handle does not turn as line comes off the reel.

Anneal – The process of heating steel and then slowly cooling it to reduce hardness, increase ductility, and reduce or eliminate internal stress in the metal. This results in the lowest hardness and maximum ductility and toughness. Steel is frequently annealed prior to cold or hot forming to enhance its formability. This is important in the manufacture of fish hooks. *See also Tempering.*

Aquatic Biology – The study of organisms that live in fresh water

Aquifer – An underground formation of permeable or unconsolidated material that holds water. *See also Spring, Appendix B.*

Aquitard – A relatively impermeable barrier of rock or other material that prevents water in an aquifer from draining downward. *Appendix B.*

Articulated Fly – A fly, usually a streamer or large baitfish imitation, which is tied on two separate metal shanks which have been joined together, allowing the fly to bend, making a longer fly which produces a more lifelike action.

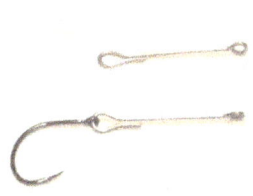

Figure 3 - Articulated Flies are tied on a shank with a hook brailing that allows the fly to bend.

Archaeology – The study of human history through artifacts.

Attractor Fly or **Attractor Pattern** – A fly which does not resemble any living insect or other bait.

Figure 4 - Attractor Pattern Royal Wulff Does not imitate any insect. Photo courtesy of Holly Flies.

Auger – A device used in ice fishing for drilling holes in the ice. May be hand powered, gas powered or battery electric powered.

Figure 5 – Lithium ion battery operated ice auger. They have largely replaced hand, gasoline and propane powered augers. Photo courtesy of The Tackle Shack, Wellsboro, PA.

B

Backcast – In fly fishing, propelling the line to the rear to provide for loading the rod to enable the cast. *See also Cast and Load. Appendix E.*

Backlash or Birds Nest – In a conventional reel when the spool spins faster than the line coming off of it, resulting in a snarl of line about the spool. Sometimes known as a Hey Dad, because that is what youngsters learning to use these casting reels say when it happens.

Backing – The thin soft line usually of 20 or 30 pound test that is attached to the reel arbor and loaded first on an empty fly reel to keep coiling of the fly line to a minimum and two, to help control fish that have run all of the fly line off the reel.

Figure 6 - Backing Comes in Bulk Spools and Single Spools. Photo courtesy of Cortland Line Company.

Baitcasting Reel - A reel mounted on top of a fishing rod that winds line around a central spool that is perpendicular to the plane of the rod, and is capable of casting a lure. *See also Level Wind Reel*

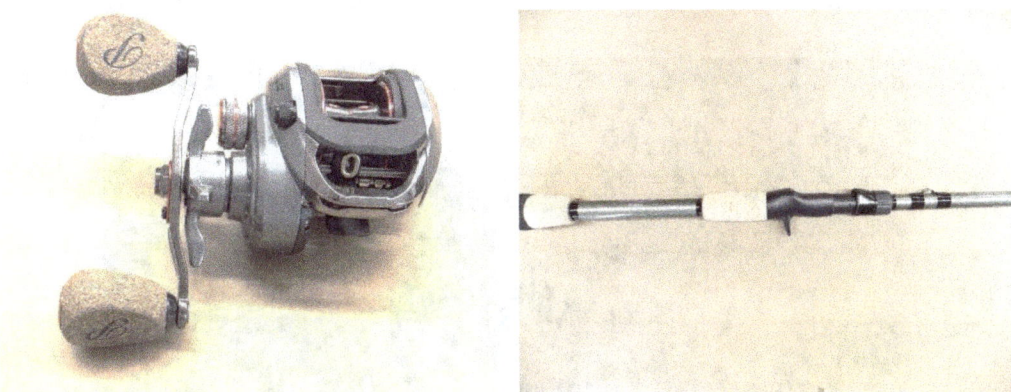

Figure 7 - Bait Casting Reels Are Hand Winding Specific. Right Hand wind is shown. Baitcasting rods have a trigger-like protrusion or finger guard located beneath the reel to facilitate casting

Baitcasting Rod – A relatively stiff fishing rod with smaller guides than a spinning rod and a finger guard located beneath the reel.

Baitfish – Any fish that is capable of being eaten by another fish. These are fish that are imitated by streamers and bucktails. Some states and provinces have legal definitions of baitfish. *See also Bucktail, Streamer, Minnow*

Baitholder Hook – A hook with one or more slices on top of the shank of the hook, designed to hold bait in place on the hook.

Figure 8 - Baitholder Hook with Barbs on Shank

Bamboo – A perennial evergreen plant of the grass family *Poaceae* and the subfamily *Bambusoideae*, which can be used for making fishing rods. Long, thin shafts of the plant can be harvested and used for fishing rods simply by attaching a line with a hook to the thin end of the shaft. Or the larger culms may be processed, machined, and glued into fishing rods. Prior to the advent of modern materials following World War II, bamboo was the preferred material for manufacturing fishing rods. It is the only plant that can be machined to thousandths of an inch tolerance consistently. This largest member of the grass family has been widely transplanted and is considered invasive in some places. Also referred to as cane, and fishing rods made from it are said to be split cane rods, referring to how the bamboo has to be split and planed.

Figure 9 – Bamboo. Larger culms are used for rod building.

Barb – A slice in the spear of the hook before the point on the inside of the bend, usually between 25º and 35º upward. *See also Beard. Appendix H*

Barbless – A hook without a slice or "Beard" in the spear of the hook. *Appendix H.*

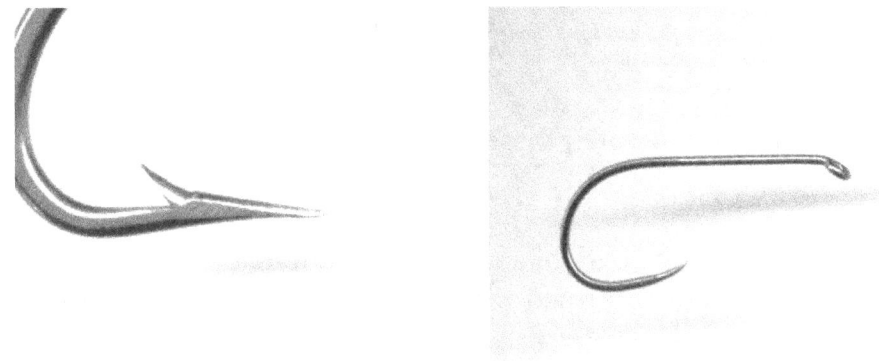

Figure 10 - Hook Barb and Barbless Hook

Bartleet Bend – A Partridge Company bend that is parabolic with a straight spear.

Bass or **Black Bass** – Fish of the genus *Micropterus* of which there are currently thirteen recognized species. They are members of the *Centrarchid* family, the sunfishes. The Smallmouth Bass, *Micropterus dolomieu*, the Spotted Bass, *Micropterus punctulatus* and the Largemouth Bass *Micropterus salmoides* are the three species most often pursued by anglers, the latter being the most popular gamefish in America. Bass may be pursued with either conventional tackle, which adds millions if not billions of dollars to the American economy each year, or fly tackle. The Smallmouth Bass is a favorite target of many warmwater fly fishing enthusiasts.

Bass or **Temperate Bass** – Members of the family *Moronidae* which consists of six species, including Striped Bass, *Morone saxatilis*, White Bass, *Morone chrysops*, White Perch, *Morone Americana*, Yellow Bass, *Morone mississippiensis*, European Seabass, *Dicentrarchus labrax*, and Spotted Seabass, *Dicentrarchus punctatus*. This family is found in salt, brackish and fresh water. Many are anadromous. Striped bass also known as Rockfish along the Atlantic Coast are migratory. Hybrids between the Striped Bass and White Bass are stocked in freshwater lakes for sportfishing and are called Wipers.

Bead, Round – A metal, glass or plastic ball with a center hole that is first attached to a hook when tying flies that will add weight to the fly, usually a nymph or streamer.

Bead, Conehead - A cone-shaped bead with a center hole placed on a usually barbless hook with the smallest diameter of the cone closest to the eye of the hook.

Bead, Slotted – A round bead with a slot instead of a center hole, used for tying jig style nymphs and streamers.

Figure 11 - Regular, Slotted and Conehead Beads. Photo courtesy of Hareline Dubbin

Beak Hook – Mostly for bait in saltwater, these are short strong hooks with a longer than average curved spear that points upward.

Beard – An antiquated name for the barb of a hook. During the hook making process a sharp chisel or knife made a cut into the unformed wire shaft creating a barb in a process known as bearding. *See also Barb. Appendix H.*

Bend – The rear section of the hook that changes the direction of the wire so that the spear is in the same vertical plane as the shank, causing both the eye and the spear to point in the same direction. *Appendix H.*

Benthic - That which is associated with the bottom of a waterbody.

Benthos – Organisms at the bottom of freshwater streams, ponds, lakes and rivers.

Big Game Fishing – Saltwater fishing for marlin, swordfish, or tuna with highly specialized equipment from a boat. Fly fishing for these big fish is possible but it requires highly specialized equipment.

Biology – The study of life.

Biome – A large naturally occurring community of flora and fauna occupying a major habitat

Black Nickel – The finish applied to hooks that give the hook a metallic black look.

Blade – A thin metal or plastic plate that is sometimes used in the front of a fly to act as a flashy fish attractant.

Figure 12 - Blades come in a variety of shapes, styles and finishes.

Blob Fly – An attractor fly tied on a short shank, heavy wire hook, consisting of a short tail and a packed body of a synthetic material resembling crystal chenille. Blob flies are not widely used in North America but are popular in the U.K and Europe.

Figure 13 - Blob Fly. Photo courtesy of The Essential Fly, Selby, North Yorkshire, England

Bluelining or Blue Lining – The practice of using topographic maps to identify small, hard to reach streams, traveling to them and fishing them, often followed by posting photos and descriptions on social media, and thus creating excessive pressure on the sometimes fragile populations of fish. *See Topographic.*

Bobber – A buoyant device, usually round or oblong that is attached to a fishing line that will move or submerge when a fish takes the bait on the hook attached to line below. *See also Strike Indicator.*

Figure 14 - Bobbers come in a variety of shapes, sizes and colors

Bobbin – A fly tying tool for holding thread.

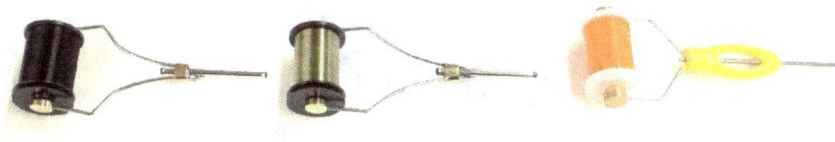

Figure 15 - Fly Tying Bobbins

Bodkin – A stout needle in a handle used in fly tying.

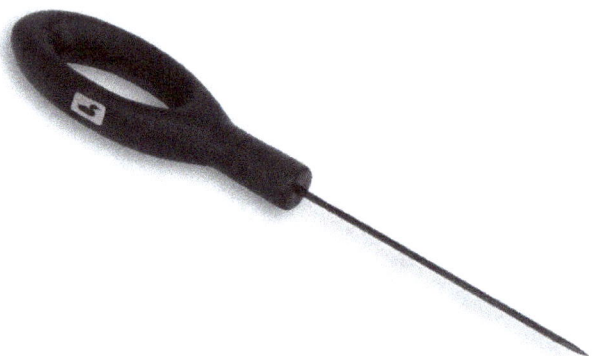

Figure 16 – Bodkin. Photo courtesy of Loon Outdoors

Bomber – A dry fly for salmon fishing constructed on a light wire upturned eye salmon hook, with a clipped deer hair body and a palmered hackle. *See also Hairbug*

Figure 17 - A Bomber with a body of clipped spun deer hair.

Booby Nymph – A nymph tied with a pair of foam eyes instead of beads, allowing the fly to float near the surface rather than the bottom. Popular in Europe, particularly the UK.

Figure 18 – Booby Nymph. Photo courtesy of The Essential Fly, Selby, North Yorkshire, England

Boot Foot – Waders or hip boots where the boot is welded to the upper.

Figure 19 - Boot foot hip boot. Most companies are shying away from boot foot hip and waders because of cost

Botany – The study of plants.

Brackish Water – Water that is slightly salty and is usually found where freshwater meets the sea.

Braid or Braided Line – Synthetic line composed of either 4 or 8 strands woven together that is used on spinning or baitcasting reels. It has a small diameter and low stretch. Its low stretch is useful when sensitive lures are being fished.

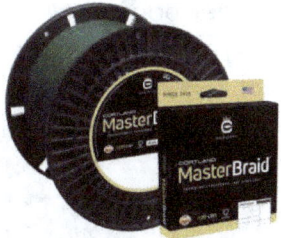

Figure 20 - Braided Line comes in bulk spools or single spools in various test strengths. Photo courtesy of Cortland Line Company

Brewster's Angle - The direction of polarization is parallel to the plane of the interface. The special angle of incidence that produces a 90 degree angle between the reflected and refracted ray is called the Brewster angle, θp. Also known as the polarization angle. It is why at certain times of the day fish and the stream bottom become clearly visible. *See Polarized Light.*

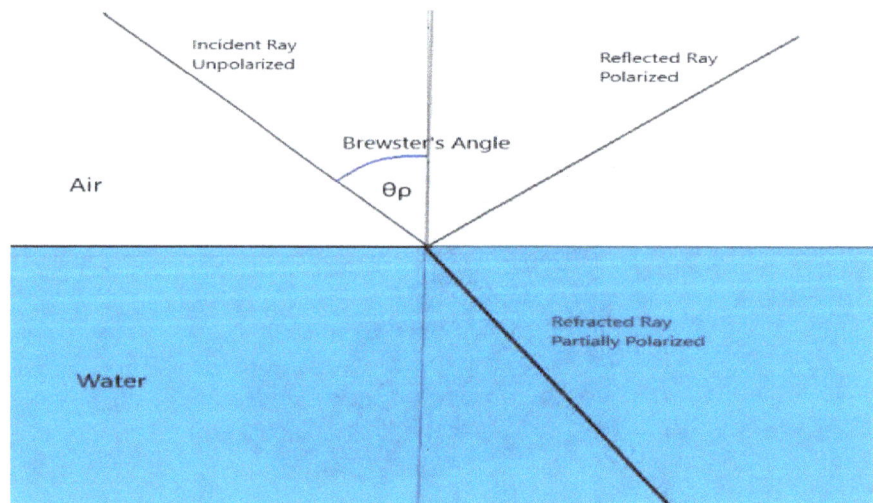

Figure 21 - Brewster's Angle or the Angle of Polarization

Bronze – 1. The finish of a hook that gives the hook a brown metallic look. 2. An alloy consisting primarily of copper and tin that was used to make hooks in antiquity.

Brush – Fur or synthetic fibers that have been twisted between two wires. The resultant wire and fiber is then used to make large streamer bodies.

Figure 22 - Brush for making bodies of streamers and nymphs. Photo courtesy of Hareline Dubbin

Bucktail – (1) A type of fly tied to generally imitate baitfish made mostly from hair. In the past two decades synthetic hair has replaced natural deer tail or other animal hair for many of the patterns. (2) Bucktail is also the hair from a deer tail used to tie flies. *See also Baitfish, Streamer*

Figure 23 - Bucktail from deer and Bucktail Streamer Photos courtesy of Hareline Dubbin and Holly Flies

Buzzer – A small midge imitation that generally consists of a bead on a hook, thread and ribbing with a slim profile to make a quick descent. Sometimes they are coated in epoxy. This is largely a European term.

Figure 24 – Buzzer. Photo courtesy of The Essential Fly, Selby, North Yorkshire, England

C

Cadastral – Relating to maps, showing property lines, government sub-divisions, buildings and other social features.

Caddis – Insects of the order *Trichoptera* that have juvenile aquatic stages and terrestrial adult stages. The adults are recognized by the 2 pair of heavy membranous wings being folded in a tent shape or inverted V over the length of the insect's body when at rest. They are not strong fliers. *Appendix J.*

Caddis Larva – The juvenile form of the caddis fly. These may be either "case builders" which construct housing of bottom sediment or organic material glued together, net building or "free swimming" which is exposed to the stream. The plural is larvae.

Captain Hamilton Bend – Found on hooks manufactured by Partridge, a near round bend with a shorter spear than average.

Carapace – The hard outer shell or exoskeleton covering the thorax of an insect, crustacean or arachnid

Carbon Fiber – Sometimes known as Graphite Fiber is a thin filament that is the basic material used to make carbon fiber cloth. That cloth is then wrapped around a mandrel to form a hollow tube that can be made into a fishing rod. Lighter than fiberglass, carbon fiber has largely replaced fiberglass in the manufacture of fishing rods.

Carbon Steel – Steel with a carbon content up to 2.1%. The carbon content determines the final properties of the metal. Usually the higher the carbon content the harder the steel. Also carbon steel rusts. This is why one should never put wet flies immediately back in the fly box.

Carlisle Hooks – 4X Long, straight eyed hooks with an offset barb, generally used for bait such as grasshoppers and crickets. This style of hook was not developed by Marinaro et. al. in Pennsylvania but rather came from Carlisle, England. *See also Marinaro Midge*

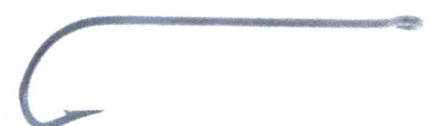

Figure 25 - Carlisle hook, thin wire with point offset to left (Kirbed)

Cast – The act of propelling a fly or lure. *Appendix E.*

Figure 26 – Fly casting. Photo courtesy of Ed Jaworowski.

Cast Net – A circular net with weights around the edges that is thrown by the angler in relatively shallow water, collecting fish and shrimp as it settles to the bottom.

Figure 27 - Using a cast net to capture bait in a brackish water sound.

Catadromous - Fish that spend most of their life in freshwater, but return to saltwater to spawn such as eels.

Catch and Release – The act of fishing in which all fish caught are released back into the water.

Catskill Dry Fly or Traditional Dry Fly – A style of dry fly that originated in the Catskill Mountains of New York in the early Twentieth Century. Theodore Gordon is largely credited with beginning this style and countless others advanced it. A Catskill dry fly has a tail that is equal the length of the shank of the hook. The wings sit upright, slightly separated and perpendicular to the hook shank. The hackle is wound around the hook shank allowing the individual hackle barbs to be perpendicular to the hook shank. The hackle is 1.5 times the distance of the gape. The hackle and tail allow the fly to ride on the surface film of the water. *Appendix I.*

Figure 28 - Catskill style Gray Fox dry fly. Photo courtesy of Holly Flies

Caudal Fin – The tail fin of a fish. *Appendix D.*

CDC – *See Cul de Cunard.*

Channel – The main bed in a river or stream. *See Thalweg.*

Channel Sinuosity – The ratio of the length of the stream to the length of the valley through which it flows.

Char – A genus of the family *Salmonidae* that are characterized by fish with light spots on a dark background and the vomer in the mouth that is toothed only on top. This genus contains the species Brook Trout *Salvelinus fontinalis*, Arctic Char *Salvelinus alpinus*, Dolly Varden *Salvelinus malma* and Lake Trout *Salvelinus namaycush*. Sea run brook trout are known as Salters, and brook trout that inhabit Lake Superior are known as Coasters. In some places of its native range the brook trout is imperiled because of habitat loss, however it has been widely dispersed in the Rocky Mountains where it is not native and is seen as a pest or an invasive species. *See also Salmonidae.*

Chemical Sharpening – After hooks are formed, tempered, and annealed acid is used to remove metal on the hook. Because the point of the hook is the thinnest part of the hook, the metal it removes creates a sharp point. *See also Mechanical Sharpening.*

Chum – Chopped fish or other material thrown overboard, or otherwise placed in the water to attract fish. Accepted in some circumstances it may be illegal in some jurisdictions.

Circle Hook – A hook with a long curvature bend, so that the point faces back toward the shank of the hook. This type of hook allows for fish to be hooked in the corner of the mouth with fewer hooks set deep in the gullet or gills.

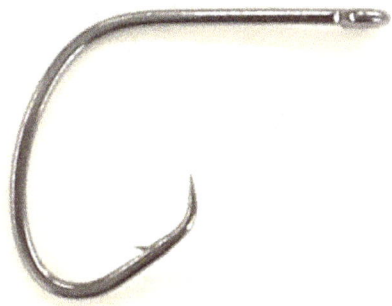

Figure 29 - Circle hook

Class - The next taxonomic group after phylum or division. Examples of classes are mammals, fish and birds.

Closed Face Spinning Reel - A reel mounted (usually) on top of a fishing rod where the spool is not visible and the line is released through a thumb button. The internal bail is closed when the handle is turned allowing for the line to be retrieved.

Figure 30 - Closed Face Spinning Reel also known as a Spincasting Reel

Closed Shank – On double hooks and treble hooks this refers to the space between the shanks that has been welded, soldered or brazed to unite the multiple shanks.

Clouser Deep Minnow – A streamer developed by Bob Clouser, constructed of real or synthetic hair and flash with dumbbell eyes that allow the streamer to sink head first as the retrieve is paused. It will be pulled along by the leader with the hook point on top. Also known as a Clouser Minnow or a Deep Minnow.

Figure 31 - Clouser Deep Minnow tied with chartreuse and white bucktail and silver flash. As Bob Clouser says, "If it ain't chartreuse it ain't no use." Photo courtesy of Holly Flies

Coaster - One of a population of brook trout that live in large water bodies. Usually when referred to the fish are in Lake Superior. *See also Char.*

Commercial Fishing – An industry devoted to harvesting large quantities of fish for sale. *See also Sport Fishing and Sustenance Fishing*

Common Carp - *Cyprinus carpio* or the common carp was originally native to Europe and Asia where it is prized as a food fish. Introduced to America it has spread rapidly and has out-competed many other native species, especially other members of the minnow family (*Cyprinidae*) to which it belongs. It is a target species for many anglers and over the past two decades has gained a following among fly anglers whose motto might be Carpe carpio! or Seize the carp!

Common Name – A name applied to an organism that is often local and confusing. The binomial nomenclature of genus and species assigns each individual species its taxonomic rank, but common names persist, as they must, for common use. The fish commonly known as walleye *Sander vitreus*, previously known as *Stizostedion vitreum*, is one fish in particular that suffers from an identity crisis. Its genus was changed a few years back, just after everyone learned to spell *Stizostedion*. In common nomenclature the fish is known as walleye, walleyed pike, yellow pike, yellow pikeperch, blue walleye, blue pike, pickerel and goggle eye. On some restaurant menus it is listed as pike. A walleye by any other name still tastes as good.

Figure 32 - Walleye Sander vitreus

Compara Dun – In 1967 Doug Swisher and Carl Richards created a dry fly that had no hackles and it floated in the surface film. First mentioned in print by Joe Brooks in his August, 1970 column in *Outdoor Life* magazine and then further popularized by Swisher's and Richard's book, *Selective Trout*, dry flies without hackle became enormously popular in the 1970s. In 1975 Al Caucci and Bob Nastasi published their monumental work, *Hatches* and introduced the world to the name Compara Dun, a hackleless dry fly with wings tied of deer hair. The flies rely on body shape and wing profile to entice the fish to strike. Today the hackleless flies are tied with a variety of materials including CDC and synthetic wing materials.

Figure 33 - CDC Compara Dun. Photo courtesy of Holly Flies

Competition – In ecology, two or more species actively seeking to exploit a resource in an ecosystem or multiples of the same species vying for the same food, habitat and spawning sites.

Competition Hook – A broad term for a barbless hook that is allowed in fly fishing competitions by the various sanctioning bodies.

Competitive Fly Fishing - The result of the competitive nature of human beings when fish are caught using fly fishing methods for prestige, sometimes monetary reward, and product endorsement, but mostly ego building.

Conservation – The act of using resources to their highest potential hopefully in a sustainable way.

Continuous Bend – A style of hook that has become popular in the 21st Century for tying insect larvae imitations, especially caddis. The bend of the hook begins immediately behind the eye with a curved shank and continues throughout the length of the hook.

Figure 34 - Continuous bend hook

Crane Fly – Aquatic insects belonging to the order *Tipulidae*. The larva is a white to brownish shaped worm. The adult resembles a large mosquito. They do not bite. They have gained some stature among fly anglers and several imitations have been developed.

Crayfish or **Crawfish** – Decapod crustaceans found in freshwater are also known as crawdads or mudbugs. They are a source of food for many fish and are often used as bait. In the past three decades Rusty Crayfish (*Orconectes rusticus*) a particularly aggressive species have spread outside of their normal range and are considered invasive. They will out-compete native crayfish and through their voracious eating habits they can drastically alter food webs. Crayfish are edible and considered fine table fare in some parts of the country.

Creel – A canvas bag or wicker basket usually worn over the shoulder by anglers to keep their catch. A 'creel limit' set forth in the regulations is how many fish you are allowed to have in your possession.

Figure 35 - Wicker Creel, Fenwick brown glass fly rod, Pflueger Medalist reel with Scientific Anglers line and a couple of trout for dinner.

Cress Bug - *Asellus aquaticus* is a freshwater crustacean that is not an insect. It is commonly found in limestone spring creeks.

Figure 36 - Cress Bug imitation. Photo courtesy of Holly Flies

Cruncher Fly – An attractor style nymph that does not usually imitate a specific insect. It is tied without a bead, and consists of a soft hackle and a hot spot of brighter material behind the hackle.

Figure 37 - Cruncher Fly. Photo courtesy of The Essential Fly, Selby, North Yorkshire, England

Cul de Cunard or **CDC** – Literally feathers from "the butt of the duck" used in fly tying for emergers and dry flies. The feathers are located near the oil glands of the duck and float. Flies tied with CDC should not be treated with silicone floatants.

Figure 38 - CDC. Do not treat flies made with CDC with silicone pastes or liquids. That will ruin the fly.

Czech Nymph – A style of fly developed to fish European rivers. It is generally an imitation of free -swimming caddis or freshwater shrimp, tied on a continuous bend hook that has added weight with a slim profile to descend rapidly through the water.

Figure 39 – Czech Nymph. Photo courtesy of The Essential Fly, Selby, North Yorkshire, England

D

Dampen - On the forward cast a fishing rod will vibrate. The more it vibrates the less accurate the cast. To lessen the vibrations the rod is designed with specific materials and built in a profile to lessen the vibrations. Lessening the vibrations is known as dampening.

Dapping – The act of placing a fly on the water without loading the rod to cast. *See also Tenkara*

Deceiver or Lefty's Deceiver – A streamer developed by Bernard "Lefty" Kreh with a slim horizontal profile, but a wide vertical profile originally tied to imitate saltwater prey, but now adapted for use in freshwater. It is tied with both feathers and bucktail.

Figure 40 - Lefty's Deceiver. Photo courtesy of Holly Flies

Denier – A unit of measure of a thread's fineness. The lower the number the finer the thread, the higher the number the thicker the thread. Pronounced Den-e-irr. *See also Thread.*

Dimictic – In lakes, twice a year (hence di) where the water mixes and the temperature at the surface is the same as the bottom. *See also Monomictic.*

Dissolved Solids - Mineral content and organic substances that dissolve in water. To measure dissolved solids a known volume of water that has passed through a 2 micron filter, then it is placed in a weighed container. The water is evaporated and the container is weighed again. Measured as mg/L

Diver – A streamer with a collar made of spun deer hair or other material that flares behind the head. The collar will make the fly dive underwater when pulled by the line and then, because of the buoyancy of the collar and head, return to the surface when the retrieve is stopped.

Figure 41 - Eric Snyder's Mad Diver. Photo courtesy of Holly Flies

Dobsonflies Alderflies Fishflies – Aquatic insects of the order Megaloptera. The adults are poor flyers and are short-lived. The larvae of the insects (known as hellgrammites) are aquatic spending 1 to 4 years growing in their aquatic habitats and are a favorite food of many fish. The hellgrammite is also a favorite bait, particularly for smallmouth bass anglers.

Dorsal Fin – The fin on the top of the fish in the middle of the back. *Appendix D*

Double Hook – A hook made from a single piece of wire with two points, and the loop eye being formed by the bend of the wire at the midpoint. This style of hook is most popular with Salmon Hooks, though there is some use of them in saltwater and warmwater lures.

Figure 42 - Brazed shank double salmon hook.

Down Eye – The eye of the hook is bent down in the vertical plane of the hook, usually around 45º but may vary between 0º and 90º.

Figure 43 - Down eye

Drag – (1) A mechanism on a fishing reel to tighten the spool to prevent line from easily coming off to help tire a fish, or loosen the spool to allow the fish to run and prevent the leader or line from breaking. (2) The movement of the water at different speeds (and sometimes wind) on the line moving the lure or bait in an unnatural manner. *See also Mend.*

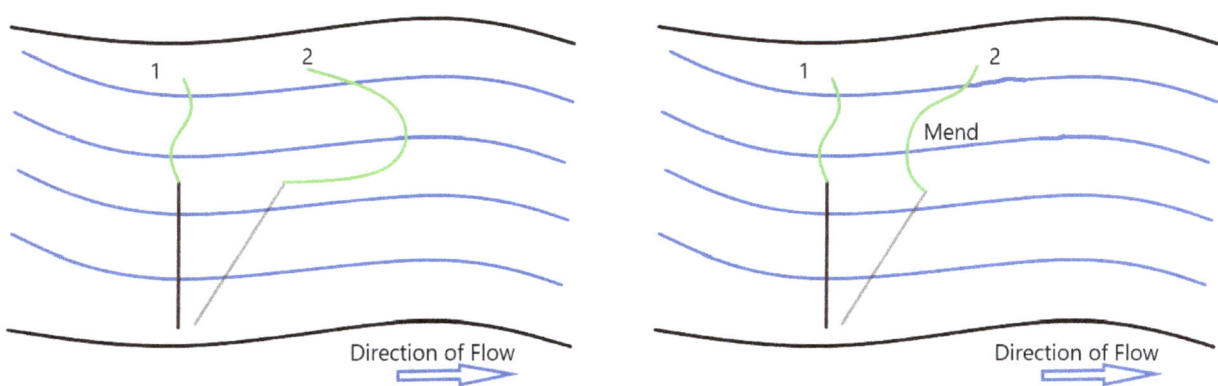

Figure 44 - Drag caused by differing water speed as it moves from 1 to 2 and corrected by a mend.

Drawing – The process of pulling metal through a die to form wire.

Drop Shotting – Attaching the weight to a blind tippet without a hook and attaching a hook above the sinker.

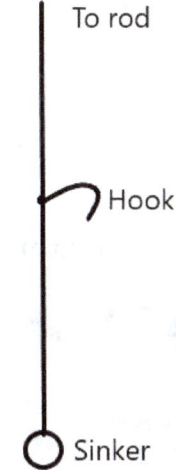

Figure 45 - Drop shot rig. Bait be it artificial or live is attached to the hook. The same setup can be used for nymph fishing

Dropper – A fly or lure tied off to the side of the main leader.

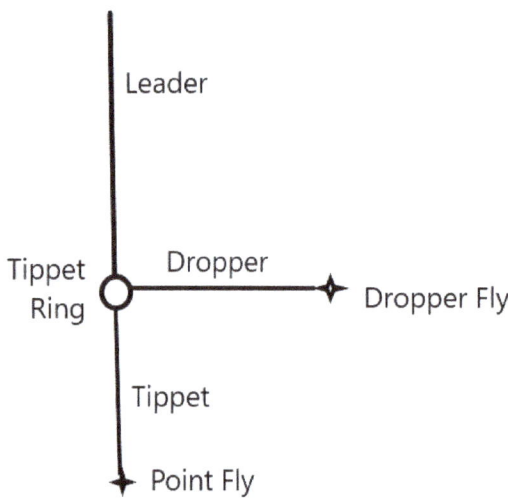

Figure 46 - A dropper may be tied to the side using a tippet ring or dropper loop.

Dry Fly – A fly tied to float on or in the surface film of water.

Figure 47 - March Brown dry fly designed to float on the surface film. Photo courtesy of Holly Flies

Dubbing – Synthetic fibers, hair, or fur that is spun around the tying thread in creation of a fly.

Figure 48 - Dubbing comes in a variety of fibers and colors. Photo courtesy of Hareline Dubbin

Dubbing Brush – A small fine-bristle brush used to tease out fibers on a tied fly's body.

Figure 49 - Dubbing brush. Photo courtesy of Loon Outdoors

Dublin Bend – A Partridge Company bend that is parabolic with a slightly upturned point. Found mostly on Partridge salmon hooks.

Dumbbell Eyes – Lead or other metal eyes in the shape of a dumbbell that are attached to a hook to give the fly weight and the appearance of eyes. See also *Clouser Deep Minnow.*

Figure 50 - Dumbbell Eyes. Photo courtesy of Hareline Dubbin

Dun – (1) A dull gray color used to describe hackle. (2) The first molt of a mayfly nymph to a winged insect. These are the mayflies that float on the water surface to dry their wings that will trigger surface feeding and is called a hatch. *See also Imago.*

Duratin – A proprietary hook finish of the Mustad Company.

E

Ecology - Modern ecology is the study of the function of how organisms relate to one another and the environment around them. From the Greek word *Oikos* meaning household or place to live *Logos* meaning study of..........

Ecosystem - An ecosystem is any area with a boundary through which the input and output of energy and materials can be measured and related to some unifying factor. The boundaries drawn around ecosystems are arbitrary and selected for the convenience of studying them. An ecosystem has both structure and function.

Egg Fly – A fly tied to imitate a fish egg.

Figure 51- Egg Fly. Photo courtesy of Holly Flies.

Egg Hook – A heavy wire hook designed for tying patterns that resemble fish eggs, in particular eggs of trout and salmon.

Figure 52 - Egg Hook

Electrofishing – The use of portable generators or batteries to power probes with electricity in the water which will stun fish for study. The fish are unharmed and released.

Figure 53 - Electrofishing a small brook trout stream.

Elk Hair Caddis or **Deer Hair Caddis** – A pattern tied to imitate mature caddis flies using elk or deer body hair to aid in flotation. Tied originally by Al Troth, a native Pennsylvanian who moved west.

Figure 54 - Elk Hair Caddis. Photo courtesy of Holly Flies

Emerger –A fly tied to resemble an insect that is swimming to the surface to molt into its next life stage. *See also Wet Fly*

Figure 55 - Soft Hackle Emerger. Photo courtesy of Holly Flies

Epilimnion – The top or upper layer of water in a stratified lake. *Appendix B.*

Equator – An imaginary line drawn around the earth that is equidistant from the north and south poles. Latitude 0° *Appendix A*

Erosion – When earth or rock is worn down and transported away by wind or water.

Estuary – The place where freshwater flowing downstream meets saltwater. Also known as bay, sound, or slough (pronounced slew). Mixing of fresh and salt water produces brackish water. Usually the area is biologically productive with active nutrient cycles.

Figure 56 - Estuary in North Carolina

Euphotic Zone – The uppermost layer of the ocean that has the highest light penetration from the sun sufficient to support photosynthesis. It is where the rate of photosynthesis exceeds the rate of respiration.

Euro Nymphing or **European Nymphing** is a fly fishing technique where an angler holds a sighter (a section of visible leader material) off the water to detect when a trout takes their weighted flies. Euro Nymph is a term applied to simple weighted nymphs. *See also High Sticking.*

Eutrophic – A lake, pond or water body rich in nutrients that provide nutrients for high vegetative growth. When the vegetation dies its decomposition uses oxygen that may cause animals in the water body to die from oxygen starvation. *See also Mesotrophic and Oligotrophic.*

Exoskeleton – The outer hard covering of insects, spiders and crustaceans made of chitin that protects the organs.

Extended Body – A fly that is constructed where a portion of the body of the fly is beyond the shank of the hook.

Figure 57 - Extended Body dry fly. Photo courtesy of Holly Flies

Extinct – The complete and irreversible removal of a species or population.

Extirpate – To completely remove an organism previously present in an ecosystem. But unlike extinction the organism may be present in some other place.

Eye – That part of a hook where the loop or circle of the wire creates a hole through which line can be threaded to connect the line to the hook. *See also Needle Eye*

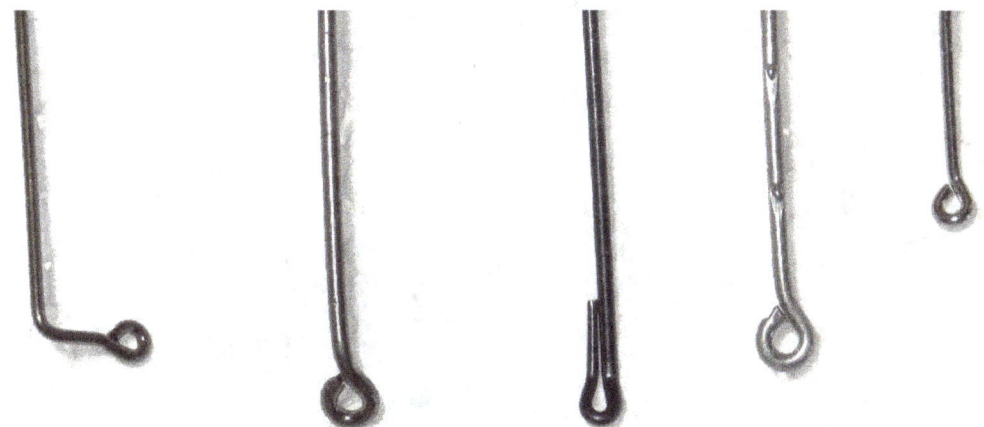

Figure 58 - Ringed jig eye, down eye, looped up eye, ring flat eye, ring up eye,

F

Fallfish – A member of the family *Cyprinidae*, fallfish *Semotilus corporalis,* are indigenous to North American waters. The fish will readily take a fly and have saved many trout anglers from having a fishless day. Fallfish are often confused with other members of the genus *Semotilus* known commonly as chubs. The reason they are so prevalent is because state and provincial fish agencies have done nothing to help them.

False Cast – In fly fishing, moving the fly line through the air to increase speed or change direction of the cast. *Appendix E.*

Family – In taxonomy the level between genus and order. Trout and salmon belong to the family *Salmonidae.* The sunfishes and black bass belong to the family *Centrarchidae.*

Fathom – A nautical unit of measurement that is equal to 6 feet (~1.8 meters).

Fecundity – The ability to reach maximum reproduction of an organism or produce many offspring.

Felt Soles – The sole of the boot or wading shoe made out of a synthetic felt material for keeping traction on stream bottoms. Felt soles are believed to be a transporter of aquatic invasive species and are now outlawed for use in some states.

Figure 59 - Felt soles on wading shoes

Ferrule – The device consisting of a male and female section that allows two pieces of a fishing rod to be joined together. Initially made of metal, most ferrules found in fishing rods today are the same carbon fiber material as the rod.

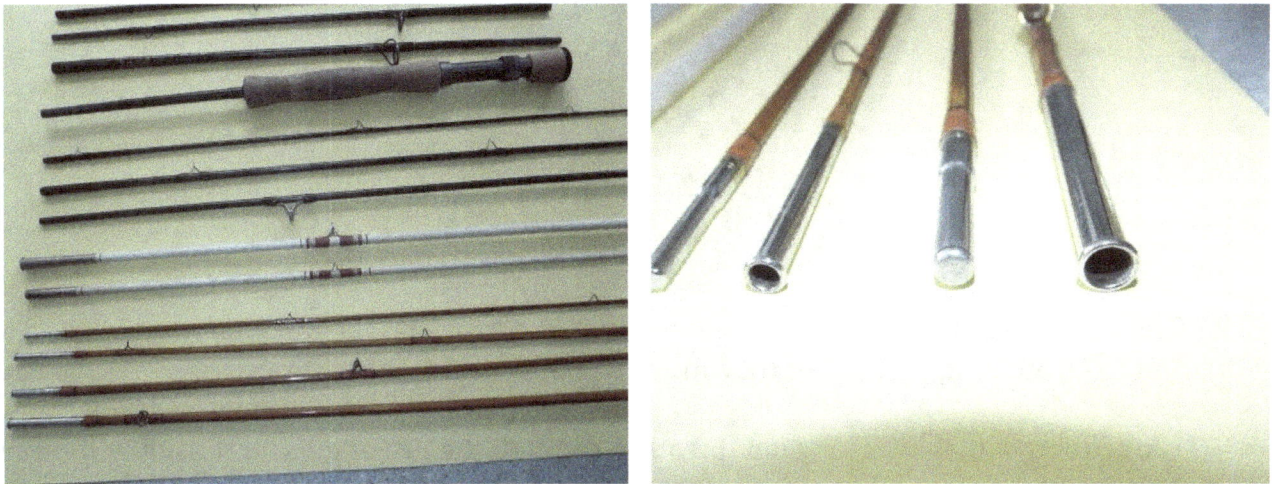

Figure 60 – Top to bottom, graphite rod ferrules, Metal ferrules on white fiberglass road and metal ferrules on a bamboo rod and close up of metal ferrules on bamboo. No matter what the ferrules are made of it is important to keep them clean.

Fiberglass – Strands of glass mixed with resin that began to be used in the production of fishing rods following the end of World War II. Early solid rods were heavy and their flexibility and strength were largely inadequate. In the 1960s techniques were developed to roll a fiberglass mat on a mandrel to form hollow tubes which could be made into fishing rods. The resultant rods were responsive and much lighter.

Fillet – The technique of separating the bones from the flesh of a fish to prepare it for consumption.

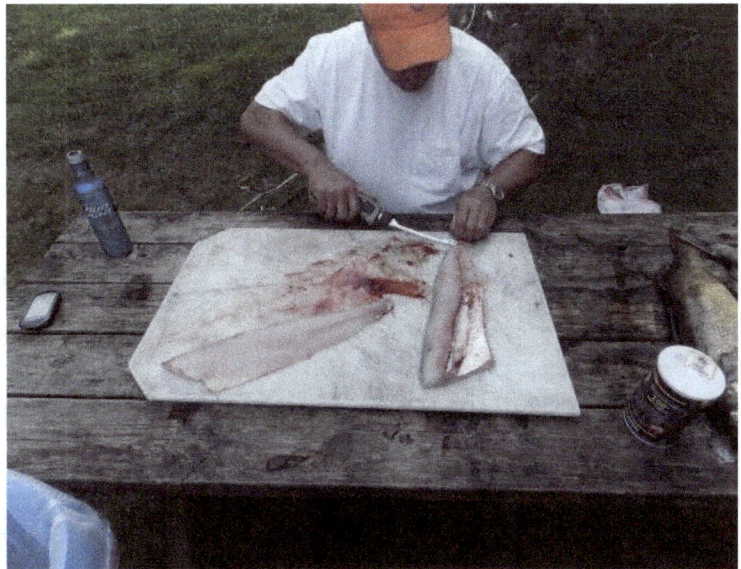

Figure 61 - Filleting fish for consumption

Fingerling – A fish usually less than one-year-old about the length of an adult human finger.

Figure 62 – Fingerling wild Brown Trout

Finish – The coating applied to hooks to prevent rust or corrosion. The finish may be applied to hooks by an electroplating process, or they may be painted, sprayed or dipped.

Fish – (1) A vertebrate that lives in water, breathes through the use of gills, and usually has fins for locomotion and stability. (2) A verb meaning to seek capture of fish.

Fish Measurement – Fish length is measured primarily by three methods, Standard Length (SL), Fork Length (FL) and Total Length (TL). Standard Length is measured from the tip of the snout with the mouth closed to back end of the last vertebra or the base of the hypural plate. Used in fisheries biology the average angler has little use for this measurement. Fork Length is measured from the tip of the snout of a fish with the mouth closed to the end of the middle of the fork of the tail. Total Length is the measurement of the fish from the snout to the end of the tail with it compressed. The angler needs to be aware of the last two types of measurement as they are used to determine if the fish may be kept or must be returned, that is the legal size limit. Girth, is measured around the widest part of the fish with the fins depressed. This measurement is used in fisheries studies and sometimes in angling records. *Appendix D.*

Fisheries Biology - The study of fish populations and their interaction with their environment. *See also Ichthyology*

Fishing Pack – A device worn around the neck and/or waist on a sling or a belt to hold accoutrements.

Figure 63 - Fishing pack. Photo courtesy of Fishpond USA

Fishing Vest - A sleeveless shirt-like garment with multiple pockets, and loops to hold fishing tackle and accessories. Lee Wulff is generally credited with its invention. Early versions were manufactured at Masland Carpets in Carlisle, Pennsylvania.

Figure 64 - Fishing vest. Photo courtesy of Fishpond USA

Flash – Plastic filaments of various colors and widths used in fly tying. Mostly made from Mylar, a product of the space program, the material is used to add color and sparkle to flies, mostly streamers.

Figure 65 - Flash. Photo courtesy of Hareline Dubbin

Flat – The front portion of the hook without an eye that has been flattened to allow for thread to be wrapped around the tippet and hook to connect the two. This is now only seen in salmon fly hooks. *See also Spade*

Flat Eye – The eye of the hook that is in line with the shank of the hook.

Figure 66 - Flat eye fish hooks.

Floatant or **Fly Floatant** – A material applied to a fly to make it resist water and float.

Figure 67 - Fly Floatants

Fluorocarbon Line – A plastic polymer line made of polyvinylidene fluoride which is more abrasion resistant, has less stretch, and is more water resistant than monofilament. It is ultraviolet light resistant and is less visible in water than monofilament. It is useful primarily as a leader material or loaded on spinning or baitcasting reels for fishing in clear water. *See also Monofilament. Appendix F.*

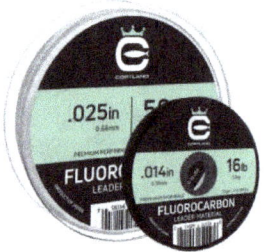

Figure 68 - Fluorocarbon leader and tippet material. Photo courtesy of Cortland Line Company

Flow Regime - Stream flow variability at a given point. It is defined by how low to how high the stream gets and how it functions at those and all points in between.

Fluvial – That which is associated with a river or stream.

Fly – 1. An insect on or near the water that fish will eat and anglers will imitate. 2. A fishing lure constructed by wrapping or otherwise securing mostly flexible material to a hook for the purpose of catching fish. *Appendix I, Appendix J.*

Figure 69 - Dry fly, wet fly, nymph

Fly Box – A box usually made of plastic, but sometimes metal or wood that is designed to hold flies.

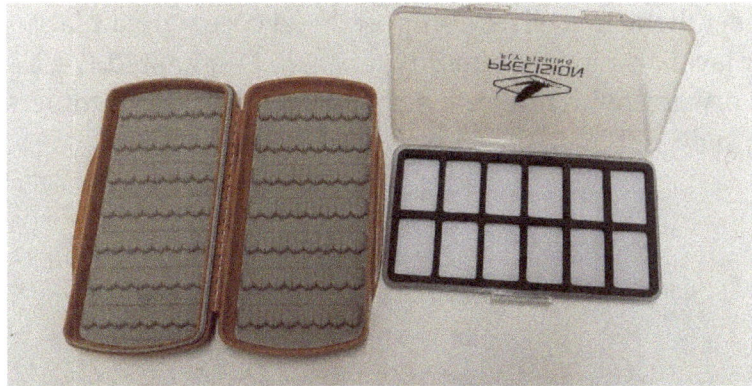

Figure 70 - Fly boxes today are generally made of plastic.

Fly Fishing - Casting an artificial lure where the weight of the line acting in concert with the energy applied to the rod carries the lure (fly) to the desired target. The fly is of insufficient weight and mass (usually) to pull the line. Though there are more than 97 families of fish worldwide and more than 350 individual species of fish in North America, three families of freshwater fish – *Salmonidae*, *Centrarchidae*, and *Esocidae* – with about fifteen species are the main targets of fly anglers. In saltwater the number of species routinely targeted by fly anglers is about ten.

Figure 71 - Fly Fishing.

Fly Line – A fishing line for casting flies originally made from silk or other natural materials, now made entirely of synthetic materials, composed of a core with a plastic outer surface. The configuration is either level or, more commonly tapered. Designated L = Level, DT = Double Taper, WF = Weight Forward, with numeric designations of lower numbers equated to lighter lines. *See also Line. Appendix E.*

Figure 72 - Fly lines come in a variety of weights, styles, tapers and colors. Photo courtesy of Cortland Line Company.

Fly Reel - A line holding device mounted on the underside of a fly rod where the line is stored and wound in the same plane as the rod. Usually they are wound in a 1:1 ratio though there are multiplier reels and spring-loaded automatic reels.

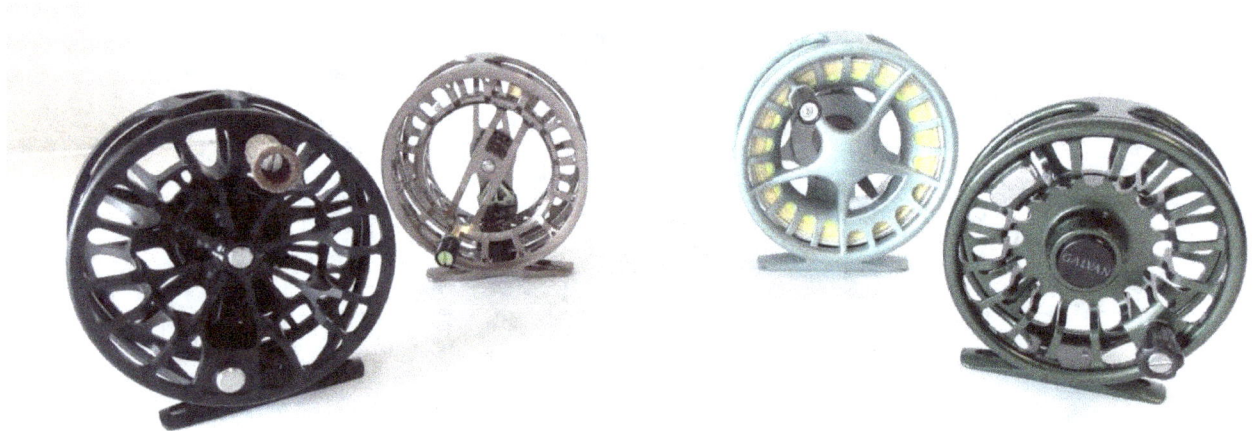

Figure 73 - Modern fly reels

Fly Rod – A flexible slender (usually) hand-held tube with guides spaced along it to allow the act of casting a fly line, and when a fish is hooked to allow the tension of a line cause by a fish at the other end to evenly transfer power and tire the fish, allowing it to be landed. Fly rods are usually more supple and longer than conventional tackle fishing rods and have the reel mounted underneath the rod at the rear of the handle.

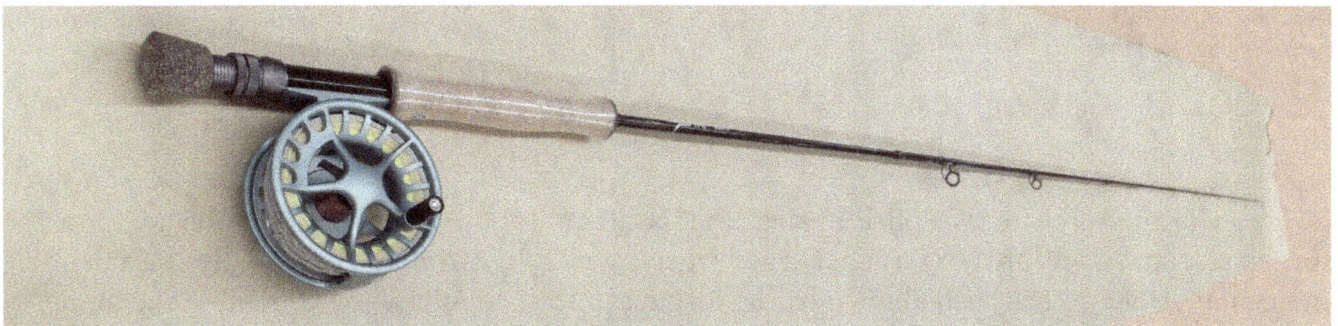
Figure 74 - Fly rod and reel

Fly Tying – The act of constructing a fly by wrapping thread and other fibers, feathers, fabrics or other materials on a hook.

Fly Tyer – A person who constructs flies. Often misspelled as fly tier. A tier is a level in a parking garage.

Figure 75 - Fly tying by a Fly Tyer and not a car in sight.

Forceps – A handheld locking clamp originally from the medical profession, used by anglers to pinch barbs on hooks, close split shot and remove hooks from fish.

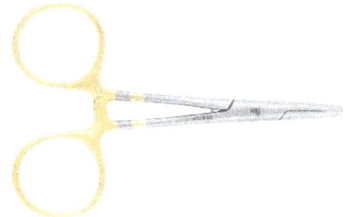

Figure 76 - Forceps have many uses to the angler. Photo courtesy of Dr. Slick.

Forge – Compressive force applied to steel, that is either at room temperature, or heated. Most forged steel products are hot forged. Forged steel is generally stronger because the internal molecular structure has been aligned by the forging process to follow the general shape of the part, in this case a fish hook. Salmon hooks are most commonly forged.

Freespooling – A method of fishing with a reel that allows the spool to revolve without any tension on the line, allowing a long line to be drifted downstream. Most often this method is seen in steelhead or salmon fishing in rivers.

Freestone Stream – Streams that flow based on the groundwater and the amount of precipitation recently fallen in their watershed. Usually they flow over sandstone or shale and/or metamorphic or igneous rocks and are very susceptible to seasonal precipitation and watershed disruption.

Figure 77 - A small Pennsylvania freestone stream. with a riffle – pool – riffle character

Freshwater – Water containing less than 1% salt.

G

Game Fish – A fish considered to be quarry for anglers. Many states, provinces and other governmental authorities have specific definitions of which fish are game fish. As a result more effort at conservation and culture is expended on game fish than other fish.

Gammarus - An amphipod crustacean genus in the family *Gammaridae*. The genus contains more than 200 described species. Also known to anglers as freshwater shrimp or scuds.

Figure 78 - A scud fly tied to imitate Gammarus

Gape or Gap – The distance between the point of the hook and the bottom of the shank of the hook.

Genotype – The individual makeup of an organism caused by its complete set of genetic material. *See also Phenotype.*

Genus - The taxonomic category that ranks above species and below family, and is denoted by a capitalized letter.

Geology – The study of rocks, minerals and the Earth's physical structure.

Ghilly or **Ghillie** or **Ghilie** – An antique term to denote an angler's helper. The term originated in the British Isles and is spelled various ways.

Gill – The organ in fish, located behind the head, responsible for extracting oxygen from water. Water enters the mouth and passes over the gills where the oxygen exchange takes place. *Appendix D.*

Gill Plate – Appendages over the gills of fish that help keep detritus from impacting the gills. *Appendix D.*

Girth – The measurement of a fish at the widest point of the fish with the fins compressed. Used in fisheries management measurements and some record fish measurements.

Glacier – An accumulation of snow and/or ice that does not melt completely from winter to winter.

Golden Rainbow Trout – A genetic mutation in one female rainbow trout in West Virginia in 1954 has led to a hybrid rainbow trout that the Pennsylvania Fish and Boat Commission have stocked since the 1960s. These fish are golden in color and stand out like a Buckeye at Beaver Stadium. They are often raised in the hatchery to significant size and once dumped into a stream they generate excitement among anglers. Sometimes called "Indicator Fish," they tend to be highly competitive with other trout and will take the best habitat. A careful angler will fish around them, usually downstream of where they are holding, as that is where the other less visible fish will be. Previously they were known as palomino trout and are now known colloquially as orange trout, goldfish, Woolworth fish and other disparaging names. While the PF&BC claims that stocking a few of them will not hurt a stream, their out-competing normal stocked fish and wild stream-bred fish, and the increasing interest in fishing the section where they are stocked does have detrimental effects by concentrating anglers who then take more than they would otherwise from a particular reach of stream and/or causing social problems with angler crowding and land abuse.

Green Weenie – Borrowing a term from the late Bob Prince, an announcer for the Pittsburgh Pirates who invented this term for something that had nothing whatsoever to do with fly fishing, the Green Weenie fly is a simple fly composed of fine lime green chenille, with a looped or trailing tail, and it may or may not have a bead head. The fly is fished sub-surface.

Figure 79 - Green Weenie. The name originally had nothing to do with fishing. Photo courtesy of Holly Flies

Guide – (1) A usually metallic ring sometimes with plastic or ceramic inserts, attached to a fishing rod meant to hold the line near the rod and transfer power from the line to the rod. (2) A licensed for profit person with qualifications (usually determined by the state, province or other political subdivision) who will take people fishing for money or other gain.

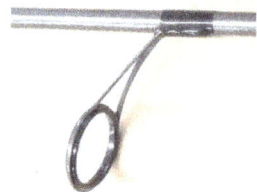

Figure 80 - Guide on a spinning rod. Steel with plastic insert.

Guide Tip – (1) The end guide on a fishing rod. It is commonly the most damaged guide on the rod and one can expect to hear cursing when damaged. (2) Customarily 20% of the total cost of a trip. Expect to hear cursing when that is not the case at the end of a trip.

Gut or Catgut – A type of cord or line made from the walls of animal intestines, usually sheep. However, catgut leaders used for fly fishing were made from silkworm caterpillar gut, the same material silk fiber was made from. They were brittle when dry and did not taper much beyond 4X. They are NOT made from cats as the name might suggest.

H

Habitat Repair – Often incorrectly referred to as Habitat Improvement, it is the human effort to reduce the impacts, or repair the errors caused by human intervention in a stream. It may involve structures being built in the stream or mitigating impacts along the waterway by the use of streamside plantings or a combination of both.

Hackle – A catch-all term describing feathers of a bird, in fly fishing usually referring to chicken feathers from the back of the neck of the bird. The hackle is composed of the stem and the barbules. Rooster feathers have stiffer barbules than hen feathers, and thus more useful for tying dry flies. Hen feathers are soft and webby and are useful for tying wet flies and nymphs.

Figure 81 – Dry Fly Rooster Hackle, prepared for fly tying.

Hackle Gauge – A measuring device used to help select the proper hackle barbule length for the fly being tied.

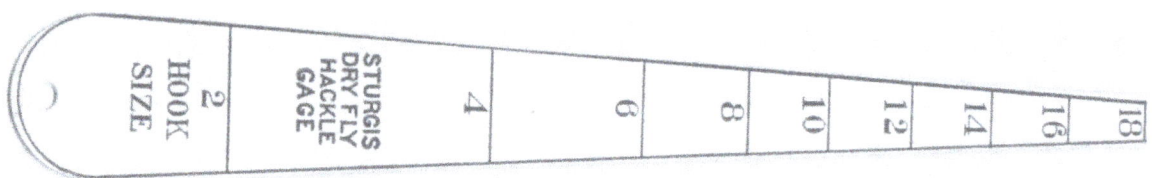

Figure 82 - Hackle gage used to size hackle. Numbers represent hook size.

Hackle Pliers – A fly tying tool used to grasp hackle while wrapping it around the hook.

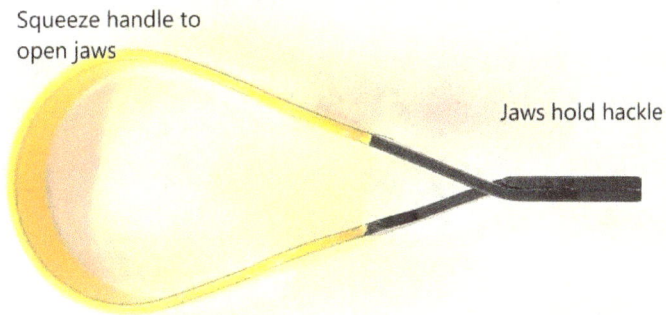

Squeeze handle to open jaws

Jaws hold hackle

Figure 83 - Hackle Pliers

Hairbug – A fly constructed of deer or elk hair which has been spun around the hook shank and trimmed to shape. Mostly these are tied for warmwater, saltwater and salmon though smaller versions such as the Irresistible or Rat Face McDougal are popular trout flies, and large Deer Hair Mice are often used for night fishing for trout. Spun hair salmon flies are known as Bombers. *See also Bomber*

Figure 84 - Hair Frog with monofilament weed guard.

Half Hitch Tool – A fly tying tool consisting of a piece of metal or plastic with a hole in the end used for creating half hitches to finish the head of a fly.

Figure 85 - Half Hitch Tool. Photo courtesy of Dr. Slick

Hard Water – Freshwater with a high amount of dissolved calcium or magnesium. Water with over 61 mg/L of calcium carbonate is classified as Hard Water See also *Soft Water*

Hardware – In this context, those fishing lures consisting of plugs, spoons, spinners, spinner baits, and so on, that are not usually fished with fly tackle.

Figure 86 – A collection of hardware.

Hatch – In aquatic insects the process when the immature forms living underwater undergo a transition to become land/airborne insects. In mayflies this is when the nymph becomes a subimago. Stoneflies crawl to above the water surface and most caddis develop underwater. *See also Dun, Imago, Mating Flight and Spinner Fall.*

Hatchery – An artificial system designed to propagate fish from egg hatching to release. *See also Nursery.*

Figure 87 – PA Fish & Boat Commission Huntsdale Fish Culture Station. Access restricted to the public to prevent the spread of invasive species, particularly New Zealand mud snails.

Haul – In fly casting pulling on the line with the free hand during the pickup and backcast to increase the speed of the line and the distance of the cast. Pulling on both a forward false cast and the backcast is a double haul. *Appendix E.*

Head – (1) The forward section of a fly line that determines how the fly line will cast. The head is responsible for loading the fly rod to make the cast and is usually 30% to 40% of the total length of the line. *See also Running Line.* (2) The front end of a fly closest to the eye, usually where the tyer ends construction of the fly.

Head Cement – Lacquer or other liquid glue used to seal threads on the head of the fly.

Figure 88 - Head cement. Photo courtesy of Loon Outdoors

Heavy Metal - A chemical element that has a relatively high density and is toxic or poisonous at low concentrations. Examples of heavy metals include mercury (Hg), cadmium (Cd), arsenic (As), chromium (Cr), thallium (Tl), and lead (Pb). Heavy metals can be found dissolved in water due to pollution from mining or industrial sources.

Heel – That part of the bend of the hook that turns the wire forward to create the spear. *Appendix H.*

High Sticking – Sometimes referred to as Tight Line Nymphing or Tight Lining is a method of fly fishing where the angler holds a fly rod as high above the water as they are able, with the line dropping vertically into the water, leading a nymph through the drift, while the line remains vertical. *See also Euro Nymphing.*

Hip Boots – Individual boots that come up to top of the thigh of an angler. They may have boot foot or stocking foot bottoms. Hip boots are useful for small streams and low water conditions.

Figure 89 - Boot foot hip boot. The boot is fused to the upper section. Photo courtesy of Frogg Toggs

Holdover – A stocked fish that has lived through a fishing season and is available during the next fishing season.

Homeothermic – An animal that maintains a near constant body temperature by using energy from food such as birds and mammals. *See also Poikilothermic.*

Hook – A piece of wire that has been shaped and pointed for the purpose of catching fish. *Appendix H.*

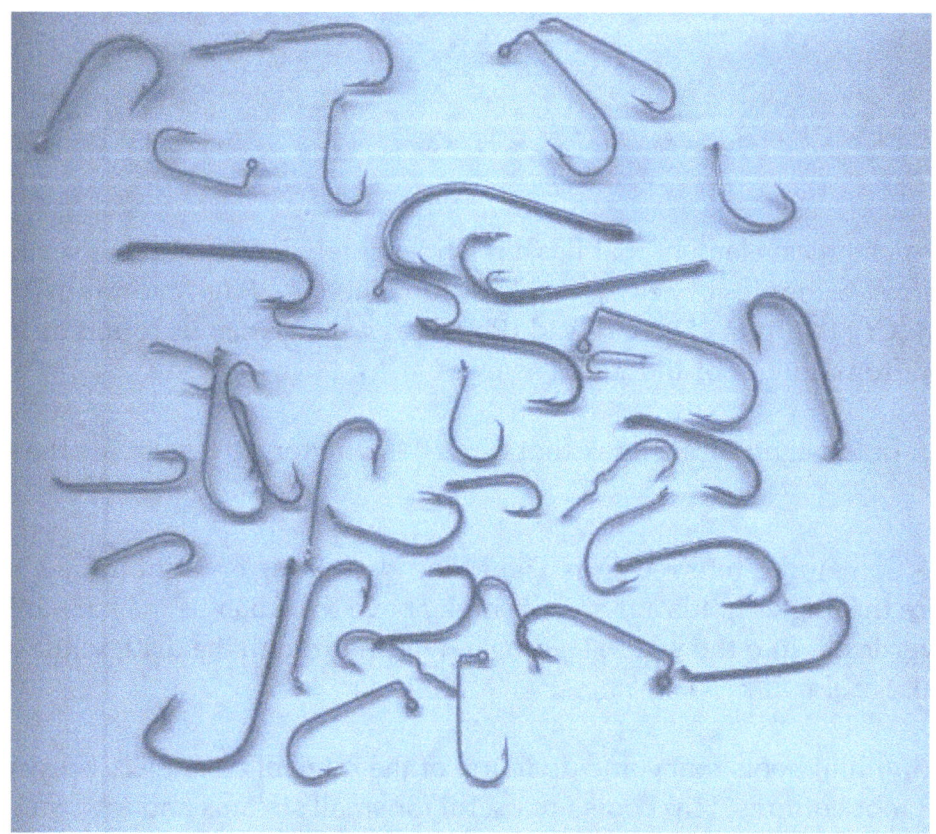

Figure 90 - Hooks come in myriad sizes, styles, and finishes.

Hybrid – In fish, used to denote the interbreeding between genuses, such as a brown trout *Salmo trutta* X brook trout *Salvelinus fontinalis* to yield a tiger trout.

Hyperbola - A symmetrical open curve formed by the intersection of a circular cone with a plane at a smaller angle with its axis than the side of the cone. Not to be confused with hyperbole which are exaggerated statements such as anglers describing their catch.

Hypolimnion – The lowest layer of water in a stratified lake, below the thermocline, often devoid of, or containing very little oxygen. *Appendix B.*

I

Ice Dot or **Ice Fly** – A hook with a molded metal shape on the shank that is usually tipped with bait for ice fishing.

Figure 91 – Ice Jig, usually tipped with an insect larva when fished. Photo courtesy of The Tackle Shack, Wellsboro, PA

Ichthyology – The study of fish. *See also Fisheries Biology*

Igneous Rocks – Those rocks formed through volcanic processes.

Imago and Subimago – Imago is the adult mayfly that is ready to mate. The subimago is the first molt of the nymph to become a winged insect that is usually duller in color than the imago. The imago is referred to by anglers as spinners and the subimago is referred to as a dun.

Intruder Fly – A large fly tied on a bare shank without a hook point, connected to a trailing hook. *See also Articulated Fly.*

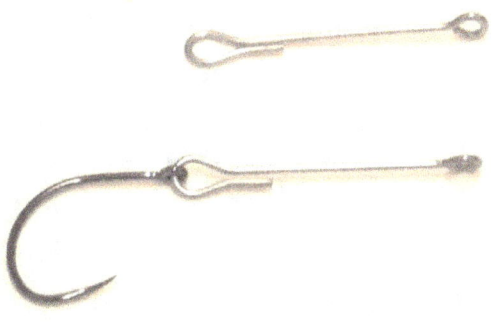

Figure 92 - Shank and hook for tying Intruder Fly

Ion – An atom or molecule in an electrically charged state because of an imbalance between electrons and protons. A cation has fewer electrons than protons and is considered positively charged. An anion has more electrons than protons and is negatively charged.

Iteroparous - In reference to fish, those species of fish which spawn more than once during their lifetime.

In Situ – Latin for "In its original place." Something is observed or work performed on in its natural context. See also *In Vivo and In Vitro*.

Invasive Species – Either a plant or animal that did not originate in a particular biome or ecosystem, that has been introduced and often out-competes similar native organisms that occupy the same niche.

Figure 93 - The author holding an invasive (but tasty) Northern Snakehead

J

J Hook – (1) A standard type of fishing hook where the spear is parallel or near parallel to the shank of the hook and forms a letter 'J'. *See Hook. Appendix H.* (2) A stream habitat device usually made of rocks in the shape of the letter 'J' that point upstream with the lower leg of the J pointing downstream. *See also Vane.*

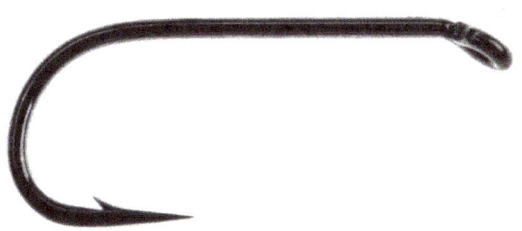

Figure 94 - A traditional J fish hook.

Jack – A sexually immature anadromous or potamodromous fish that returns to its natal stream before it is ready to spawn.

Jig – (1) A lure which has a lead or other metal head molded on the hook. (2) The act of fishing in which the up and down motion of the rod tip causes the lure or bait to move up and down.

Figure 95 - Jig tied with estaz body and marabou tail.

Jig Hook – A type of hook with an extended eye in the vertical plane of the hook. Usually a lure constructed on this hook will have weight near the eye, causing the hook to be pulled through the water with the bend and point riding up. The eye may be bent in an angle from 20º to 90º.

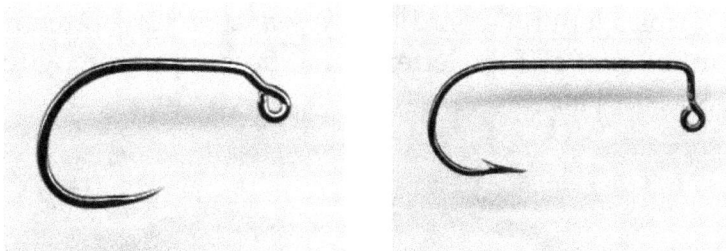

Figure 96 - Jig hooks. Fly hook on the left, conventional jig hook right.

Juvenile – When referring to fish a non-sexually mature fish, often called parr. Identified in Salmonids by lateral bars on the sides. When referring to the youngsters downstream of you who are throwing rocks and sticks in the water, the word is usually followed by delinquent and preceded by words that are unprintable here.

K

Kahle Hook – A hook with a different shape from a conventional hook. It has a continuous bend with the spear slightly offset but in the same horizontal plane as the eye. The wide gap and twisted shank are believed to cause deep hook sets and higher mortality in fish caught with these style hooks.

Figure 97 - Kahle hook.

Keel Hook – A hook designed so that the point of the hook rides up reducing the chance of snags. The original Keel Hook had a ring eye with a short shank then a 45° downward bend. The downward section would continue for the distance of the gape of the hook. The hook shank continues straight to the round bend of the hook.

Figure 98 - Keel Hook., Mustad 79666. No longer available.

Kingdom - One of the primary divisions into which natural objects are commonly classified, plant or animal.

Kink Shank – Sometimes referred to as Kinked Shank is a hook with a vertical "S" bend or inverted "V" in the middle of the shank of the hook. This expansion of the vertical plane is to allow for better adherence of popper bodies to the hook.

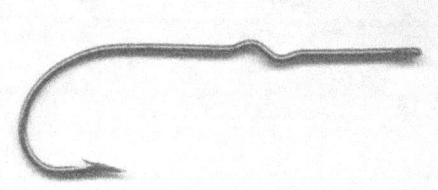

Figure 99 - Kink Shank hook for tying poppers.

Kirb – *See Offset*

Klinkhamer or **Klinkhammer Hook** – Spelled with either one or two m's, is a continuous bend fine wire hook for tying dry flies and emergers that leave the body of the fly below the surface film of the water. This method of fly tying was originally developed by Hans van Klinken.

Knot – The twisting of line in which it becomes intertwined with itself or another line. Anglers use several types of knots for connecting lines to each other, leaders to lines, or lures to leaders.

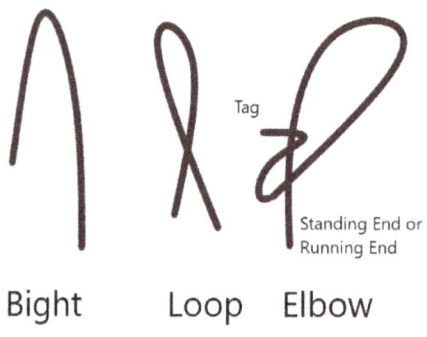

Bight   Loop   Elbow

Figure 100 - Basic components of all knots

Kype – The hooked lower jaw of male salmonids that are ready to spawn.

Figure 101 - A mature male salmon with the beginning of a kype jaw.

L

Lacustrine – Relating to lakes

Landlocked – When relating to fish, those fish populations that have been cut off from migration to the sea, even though other populations of the same species migrate.

Lateral Line – A system of sensory organs running the length of the body of a fish that are responsible for detecting movement and vibration, orientation in the water, spatial relations to other fish and possibly taste and smell. *Appendix D.*

Latitude – The distance north or south of the Equator measured in degrees, minutes and seconds. *See also Longitude. Appendix A.*

Leader –The section of line between the actual line and the hook. The end section of the leader that attaches to the hook is the Tippet. The leader's purpose in fly fishing is to allow the fly to gently straighten and present the fly at the end of the cast. Leaders are usually made of nylon or fluorocarbon and diminish in size from the line to the tippet. These are known as Tapered Leaders. Leaders that do not diminish in size are known as Level Leaders. In bait fishing or lure casting usually a level leader is used and is primarily for invisibility. *See also Tippet and Line*

Figure 102 - Leaders may come pre-made, either single piece or knotted, or anglers may construct their own

Lentic – Used to describe still freshwater habitat *See also Lotic*

Level Wind Reel - A reel used for casting or trolling, mounted perpendicular to and on top of a rod, that winds in or spools out line with the spool of the reel turning on both the cast and retrieve. *See also Baitcasting Reel.*

Limestone – A carbonate sedimentary rock formed when calcium carbonate is precipitated out of water, and the precipitate is then subjected to other geologic forces to form rock.

Limestone Stream – A low gradient stream with near constant temperature and flow that emanates from springs in limestone bedrock with water that has a high pH and alkalinity.

Figure 103 - Letort Spring Run, a quintessential limestone spring creek.

Limerick Bend – A modified Sproat Bend that changes its shape radically at the heel of the hook to produce a spear that is parallel to the shank of the hook from the heel forward.

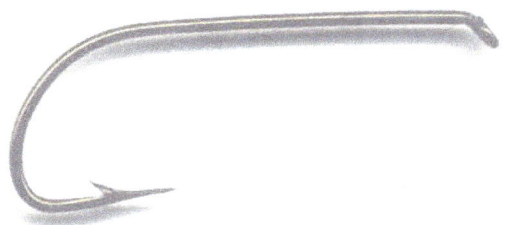

Figure 104 - Limerick Bend characterized by flat spear parallel to shank.

Limnetic Zone – Open water that admits light to make a lake productive. *Appendix B.*

Limnology – The study of physical, chemical and biological characteristics of freshwater.

Littoral – On or near the shore of the sea, usually described as the area between low tide and high tide in the ocean. In freshwater the transition from dry land to lake, usually with rooted aquatic plants. *Appendix B.*

Line – Thin pliable string (usually of synthetic composition) that is cast by an angler. In fly fishing the line is cast and the fly goes along for the ride as opposed to hardware fishing where the lure is cast and pulls the line from the reel. *Appendix E.*

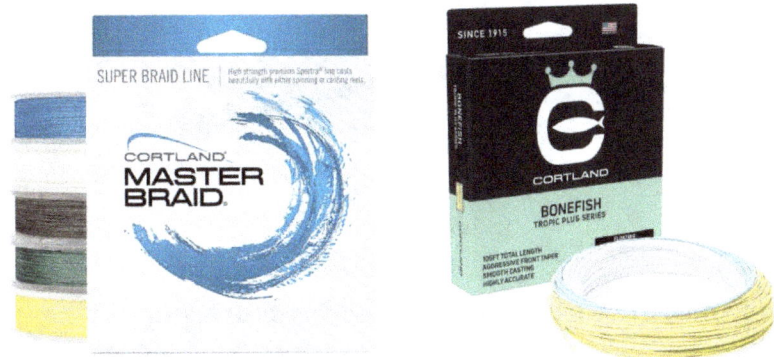

Figure 105 - Fishing lines come in many types and many are designed for specific purposes. Photos courtesy of Cortland Line Company

Live Bait – Any bait that is living or once was that is affixed to a hook and used to catch fish. Includes, worms, small fish, crustaceans, insects, fish eggs, and even paste baits in some locations. *See also Paste Bait.*

Load – In fly fishing, imparting energy to a fly rod on the back cast, causing the line to travel in the reverse direction of the intended cast, bending the rod to provide optimum energy for the forward cast. In baitcasting or spinning the weight of the lure provides the weight to bend or load the rod to help propel the lure. *See also Head and Running Line. Appendix E.*

Longitude – The distance east or west of the Prime Meridian located at Greenwich, England and measured in degrees, minutes and seconds. *See also Latitude. Appendix A.*

Loop – In fly fishing, the bend in the line between the tip of the rod and fly. Usually, the tighter the loop the more efficient the cast. When the fly crosses the line it is said to be a closed loop and makes for an ineffective cast. *Appendix E.*

Figure 106 - Tight Loop is the desired outcome of a fly cast. A closed loop will result in a bad cast and wind knots in the leader.

Looped, Up Eye – The eye of a hook made by looping the wire back partially along the shank. This type of eye is most often seen in Salmon Hooks.

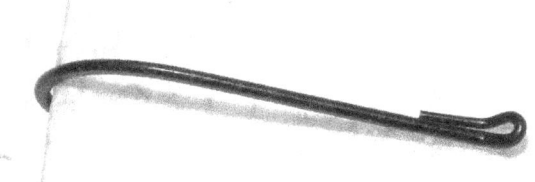

Figure 107 – Looped, Up Eye of a salmon fly hook.

Lotic – Used to describe moving freshwater habitats. *See also Lentic*

Lure – A generic term for an artificial creation designed to catch fish.

M

Macroinvertebrate – An animal without a backbone that can be seen with the unaided eye. Most commonly an insect, crustacean or worm. Aquatic macroinvertebrates are those animals that live all or the majority of their lives in water. They are used as indicators of water quality.

Marabou – Feathers originally collected from Marabou Storks in Africa, but now almost all marabou feathers come from domesticated turkeys. A marabou feather has fine fibers and can be dyed to almost any color. It moves enticingly when tied in a fly or a jig.

Figure 108 - Marabou feather.

Marinaro Midge Hook - A fine wire hook with a slightly offset bend designed by Vince Marinaro and manufactured by Partridge for tying small dry flies for use in Southcentral Pennsylvania's limestone streams. The hooks in sizes 24 – 32 are now out of production. *See also Midge Hook.*

Marine Biology – The study of life that lives in saltwater.

Marsh – A low lying area with water present but with few if any trees, where vegetation consists mainly of grasses and herbaceous plants. *See also Swamp.*

Figure 109 - Salt marsh, Swansboro, NC

Mating Flight – After final molting the sometimes complex flight patterns of male aquatic insects to attract females for mating. *See also Hatch and Spinner Fall.*

Mayfly – Insects belonging to the order *Ephemeroptera* that spend their juvenile life underwater. The immature stages have abdominal gills with usually three tails. When hatching they undergo incomplete metamorphosis, unique to the order and have a winged terrestrial immature state, before undergoing a final molt in which the adult has four clear upright wings, long tails and no mouth.. There are more than 3,000 species of mayflies. *Appendix J.*

Figure 110 - Slate Drake mayfly

Mechanical Sharpening – Using abrasives or metal working tools to sharpen the point of a hook. This process has largely gone out of favor, especially on smaller fly hooks, being replaced by chemical sharpening. *See also Chemical Sharpening.*

Mend – In fly fishing, using the fly rod to pick up sag of the fly line from the water to help the fly achieve a more natural or longer drift.

Mesotrophic – A lake that has moderate amounts of nutrients and aquatic life. *See also Oligotrophic and Eutrophic*

Metamorphic Rock – Igneous or sedimentary rock that has had its structure altered by geologic processes

Midge – A small two-winged aquatic insect of the *Chironomidae* family. There are over 10,000 species. They are not mosquitos.

Midge Hook – A small hook for tying imitations of members of the *Chironomid* family. Usually in sizes 20 to 26. *See also Marinaro Midge Hook.*

Minnow – A fish that belongs to the family *Cyprinidae* that contains over 3,000 species ranging in size from half an inch to several feet long. Also minnow is used as a generic term to denote small fish or bait fish. Not all small fish are minnows and not all minnows are small fish. Frequently imitated by fly tyers and bait manufacturers for lures. *See also Baitfish.*

Monofilament – A line made from a single fiber of nylon or some other type of plastic. Stren, introduced by the DuPont Company in the late 1950s is generally considered the first successful monofilament fishing line. Monofilament may be used for line on spinning reels, baitcasting reels, and even fly reels in some situations. Its primary use in fly fishing is for leader material. *Appendix F.*

Figure 111 - Monofilament comes in a variety of sizes and styles.

Monomictic – In lakes where once a year the water mixes and the temperature at the surface of the lake is the same as at the bottom of the lake. *See also Dimictic*

Mop Fly – A wet fly or nymph (it is hard to say which) constructed from the soft segments found of floor mops, or from the segments found on some indoor rugs. The segment is generally bound to the hook only at one end, with a collar in front. The back end of the segment is allowed to hang free. It may be tied with or without a bead head.

Figure 112 - Mop Fly

Morphology – A branch of biology dealing with the study of the organism's form and structural features.

Morphometry - The process of measuring the external shape and dimensions of landforms, living organisms, or other objects. Traditional morphometrics analyzes lengths, widths, masses, angles, ratios and areas. In general, traditional morphometric data are measurements of size.

Muddler Minnow – A streamer with a spun deer hair head. Invented by Don Gapen of Minnesota in the 1930s, the streamer is designed to imitate a sculpin. *See also Sculpin.*

Figure 113 - Muddler Minnow. Photo courtesy of Holly Flies

N

Native Fish – Fish that were originally found in the water before human intervention. *See also Wild Fish and Stocked Fish*

Figure 114 - A native stream-bred brook trout.

Nautical Mile - 1,852 meters. 1 Nautical mile per hour equals 1 Knot.

Needle Eye – An eye made by punching through the shank of the hook, as opposed to bending the wire around a mandrel to form an eye. A needle eye allows a thinner horizontal profile, especially in large saltwater hooks.

Needle Point – (1). A round cone-shaped point on the hook with no angled sides. (2). A barbless hook popularized by Seth Green in the 1870's where the hooks were made from needles without being bearded.

Net or Landing Net – A teardrop or circular framed net of varying size with a flexible soft mesh or rubber basket used to land fish.

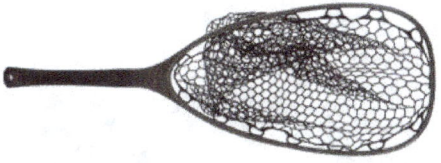

Figure 115 - Landing Net. Photo courtesy of Fishpond USA

Niche – The match of a species to its environment and how the environment affects the species, and how the species affects the environment.

Nickel Finish – A dark gray metallic finish applied to hooks by an electroplating process mostly seen in saltwater hooks.

Nipper – A device for cutting leader and tippet.

Figure 116 - Nipper with eye cleaning needle and lanyard to attach to a retractor. Photo courtesy of Dr. Slick.

Nursery – An artificial system designed to raise young fish to adult size suitable for release. *See also Hatchery*

Figure 117 – The Yellow Breeches Anglers and Conservation Association, Alma Lee Thorton Cooperative Trout Nursery, Boiling Springs, Pennsylvania. The largest cooperative nursery in Pennsylvania. Trout are raised from fingerling to adult.

Nymph – An immature form of an aquatic insect found beneath the surface of the water, most commonly on or near the bottom or the imitation tied to represent the real thing.

Figure 118 - Pheasant Tail Nymph. Imitates many different species of mayflies. Photo courtesy of Holly Flies.

O

Octopus Hook – A short shank hook, usually with an up eye and round bend, with an upturned spear that is not quite as severe a bend as a Circle Hook.

Figure 119 - Octopus hook.

Offset Point – A bend in the hook at the heel that moves the point of the hook outside the vertical plane of the hook. Looking down on a hook from the top while holding the bend, a point moved to the left is often referred to as a kirb or kirbed point. A point that is deflected to the right is called reversed. In this book to avoid confusion all hooks with points that extend outside the vertical plane of the hook are referred to as offset.

Oligotrophic – A nutrient poor lake with minimal aquatic life *See also Eutrophic and Mesotrophic*

Open Faced Spinning Reel – Most often referred to as a Spinning Reel is a fishing reel mounted below a fishing rod with a visible spool of line that does not move. A bail on the reel opens to allow the line to cast and the bail closes and rotates about the spool (driven by a handle and gears) to retrieve the line. *See Spin Fishing.*

Figure 120 - Open Faced Spinning Reel or more commonly Spinning Reel

O'Shaughnessy Bend – A forged hook made of heavy wire with a modified sproat bend, somewhere between a round bend and a sproat bend. Mostly used for saltwater and not commonly seen in hooks designed for freshwater flies. Many of these hooks have an offset point.

Figure 121 - O'Shaughnessy Bend

Order – The taxonomic rank between family and class. Trout belong to the order *Salmoniformes*. Sunfish belong to the order *Perciformes*.

Ought or **0** – A size measurement used to describe the size of the hooks larger than a size 1 hook. Originally described as a series of zeroes, it is now mostly expressed as a numeral, followed by a slash and a zero.

0	1/0	One ought	Smallest
00	2/0	Two ought	
000	3/0	Three ought	
0000	4/0	Four ought	
00000	5/0	Five ought	Largest

The numbers go as far as 12/0 for some saltwater hooks. As the number of "oughts" gets higher the hook size grows larger.

P

Paleontology – The study of plant and animal fossils

Parabola - A curve where any point is at an equal distance from: a fixed point (the focus), and a fixed straight line (the directrix). In short, a section of an ellipse. In this context, the shape of the bend of many types of hooks, particularly Sproat Bends.

Parachute Dry Fly – A dry fly where the hackle is wound above the shank of the hook around a post of hair or synthetic fiber. A Klinkhamer fly is one type of a parachute fly.

Figure 122 - Parachute Dry Fly with hackle wound on a post above shank rather than around it. Photo courtesy of Holly Flies.

Parr – A juvenile trout or salmon not yet old enough to spawn, identifiable by dark longitudinal stripes or splotches on its side.

Figure 123 – A parr brook trout. Photo courtesy of Gary Leone

Paste Bait – Manufactured bait with a putty-like consistency that has odorants or other fish attracting features that is molded around a hook. *See also Live Bait*

Pectoral Fins – A pair of fins located behind the gills on the side of the fish. *Appendix D*

Pelagic – Relating to the open ocean, not near the bottom or the shore.

Pelvic Fins – A pair of fins located on the bottom of a fish ahead of the vent behind the gills. *Appendix D.*

Perfect Bend – *See Round Bend, also Viking Bend.*

Personal Floatation Device or **PFD** – A vest, jacket or other device worn by anglers, boaters and other water users to keep them afloat in the event of their falling into the water. There are government requirements as to what kinds and how many should be on any particular watercraft and when and how they should be worn.

Figure 124 - PFDs come in 5 types. Select the one that is best for your needs. From OSHA

pH – A scale denoting the potential of hydrogen to specify the acidity of an aqueous solution with 7.0 being neutral, less than 7 becoming more acidic as the number decreases to 0, and more basic as the number increases to a maximum of 14.

Phenotype – A set of observable characteristics of an individual resulting from its interaction with its environment and its genotype. *See also Genotype.*

Philopatry - The tendency of an organism to stay in one place, often the place they were born.

Phylogeny - The history of the evolution of a species or group, especially in reference to lines of descent and relationships among broad groups of organisms. Phylogenetics in biology is a part of the systematic address of the inference of the evolutionary history and relationships among or within groups of organisms. A phylogenetic tree is a branching diagram or a tree showing the evolutionary relationships among various biological species or other entities based upon similarities and differences in their physical or genetic characteristics.

Photosynthesis – The chemical reaction of plants to capture light energy and using carbon dioxide and water synthesize food and give off free oxygen. Described by the equation:

$$6CO_2 + 6H_2O \rightarrow C_6H_{12}O_6 + 6O_2$$

Phylogenetic – Relating to evolutionary development of a species or group or organisms.

Phylum – The taxonomic class below kingdom and above class. Fish belong to the phylum Chordata those animals with a central nervous system, bilateral symmetry, a circulatory system and segmentation

Phytoplankton - Microscopic aquatic plants that are non-rooted

Pike – Members of the family *Esocidae*. The genus contains seven species. Of interest to anglers are the northern pike (*Esox luciens*), American pickerel (*Esox americanus*) of which there are two subspecies, redfin pickerel and grass pickerel, chain pickerel (*Esox niger*) and muskellunge (*Esox masquinongy*). The Tiger Muskie is usually sterile, hybrid offspring of the true muskellunge (Esox masquinongy) and the northern pike (Esox lucius). Because of their fast growth rate they are often raised in hatcheries and stocked. The members of the pike family are top predators and highly sought after gamefish. In the past decade the fish have become targeted by fly anglers, using heavy rods, large flies and bite tippets.

Figure 125 - Northern Pike

Piscicide - A poison designed especially to kill fish. Sometimes used in fisheries management to rid a body of water of undesirable species.

Piscivore – A fish eating animal. An animal that eats fish.

Plankton – Microscopic and near microscopically sized plants and animals that drift about in open water at the mercy of wind and currents. They are a substantial part of the food web.

Plastic Bait or Soft Plastic – Flexible molded bait that may or may not resemble living organisms, that comes in various colors and may be scented. The baits come without hooks, and are rigged in various ways on a multitude of various hook shapes and sizes and usually fished with a baitcasting outfit or a spinning outfit. In recent years fly tyers have discovered small plastic worms that they lash to a hook and call Squirmy Wormies.

Figure 126 - Soft Plastic baits.

Poikilothermic – An animal with a body temperature of the surrounding environment. Often referred to as coldblooded animals. Examples are fish, insects, amphibians and reptiles. *See also Homeothermic.*

Point – The sharp end of the wire on the bottom of the vertical plane of the hook.

Polarized Light – Light in which the waves vibrate in only one direction. *See Brewster's Angle.* By causing the light to be polarized by using filters such as sunglasses, and whether or not the angler is aware of it because of Brewster's Angle, an angler can more easily see fish in water because the surface glare has been eliminated.

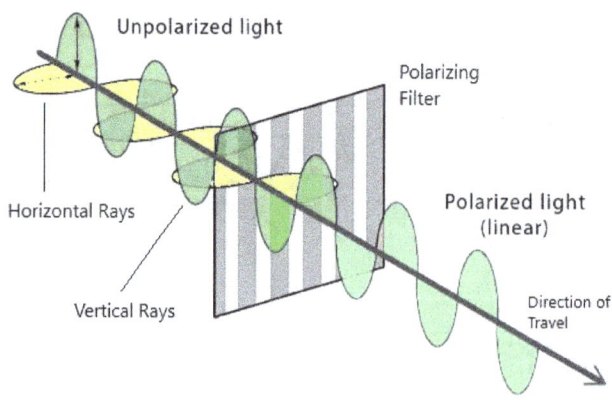

Figure 127 - Polarization is light moving in only one direction. In this example horizontal light waves cause glare and are removed by a polarizing filter.

Popper – A fly constructed of a cork, balsa wood, or foam body designed to be fished on the water surface. The flies are so designed to make noises that attract fish. Some poppers have more subtle action than others.

Figure 128 - Hard body foam poppers.

Population – The major functional unit of an ecosystem. Multiple individuals of the same species.

Polymorphism – The occurrence of two or more clearly different forms of the same organism.

Polyphenic – The occurrence of several different phenotypes of the same organism caused by the environmental rather than genetic conditions. For example a light colored trout that lived all its life over a sand bottom stream in a meadow and a dark deeply colored trout that lived its life underneath hemlocks next to rocks are both the same Genus and species.

Potamodromous - Fish that are born in tributary streams migrate to a larger body of freshwater and return to the tributary stream to spawn, such as Lake Erie steelhead.

Prime Meridian – An arbitrary line of longitude that runs from pole to pole with a reference of 0°. The world standard is a line that runs from the North Pole to the South Pole through Greenwich, England. *See also Longitude. Appendix A*

Profile – The shape of the fish hook, also referred to as the hook profile.

Profundal Zone – A deep zone in a lake where light does not penetrate

Pupa, plural **Pupae** – The life stage of some insects between immature larva and adult. In fishing, caddis flies are among the most common insects anglers encounter that have a complete metamorphosis.

Q

Qualitative – Relating to measuring the quality of something, e.g. Fair, Poor, Good or Absent, Present, Common

Quantitative – Measuring something in a quantity, e.g. mg/L or number per square meter.

R

Redd – A spawning place for fish, usually constructed by the fish before spawning.

Figure 129 – Salmon spawning on the Manistee River, Michigan

Reel Seat – That part of the fishing rod handle that through the use of moveable rings holds a reel on a rod. Downlocking means the rings are closer to the handle and tighten the reel by moving toward the butt end i.e. away from the tip. An uplocking reel seat has rings located at the back end of the rod and move toward the handle and the tip to secure the reel to the rod.

Reflection – Light waves being bounced off a shiny surface. To the angler this is what produces glare on the water. *See also Refraction.*

Refraction – When light passes from one medium to another, air to water for example the direction of a wave will change which is caused by its change in speed of the light wave. *See also Reflection.*

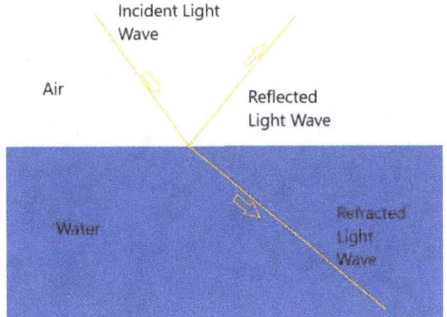

Figure 130 - Reflected and Refracted Light

Retainer Bend – A hook bend that has two distinct rounded corners, sometimes referred to in older literature as a Sneck Bend though in modern usage a Sneck Bend is distinct. *See also Sneck Bend*

Retractor – A spring loaded cord that is pinned or clipped to an angler's vest or shirt to which any number of tools may be attached. Some are large enough to hold a landing net secure. Sometimes called zingers by anglers.

Figure 131 - Retractor. Photo courtesy of Dr. Slick

Retrieve – (1) Describing reels that says which hand will wind the handle, such as "Right hand Retrieve" (2) The method of taking in line that determines the action of the lure or bait.

Reversed Point – *See Offset Point*

Ring Eye – The eye of the hook that is made by bending the wire back to the axis of the shank of the hook rather than looping it. It may be flat, down, up, or vertical.

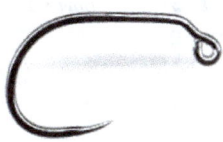

Figure 132 - A vertical ring eye nymph hook.

Riprap – Large rocks used for erosion control or habitat devices.

Figure 133 - Riprap used for bank stabilization

Riparian – Relating to the banks of a river or stream

Riparian Buffer – A vegetated strip along a river or stream. Riparian buffers slow down erosion and otherwise help reduce pollution entering a stream. The strip can be of any width, though wider is considered better in terms of resource protection.

Figure 134 - Trees and herbaceous growth act as a riparian buffer

River – A generic term, not defined by size, of a naturally flowing watercourse.

Figure 135 - Rivers may be any size. To paraphrase, "They don't say nuthin but they must know sumthin. They just keep rollin along.

Rod, Fishing - A flexible slender (usually) hand-held tube with guides spaced along it to allow for casting the lure or bait, and to evenly transfer the tension of a line caused by a fish at the other end of the line. The action of a rod is determined by its profile, based on the mandrel it was wrapped on, the materials used in construction and sometimes the guide spacing and wrapping. A slow rod is limber. A fast rod is stiff.

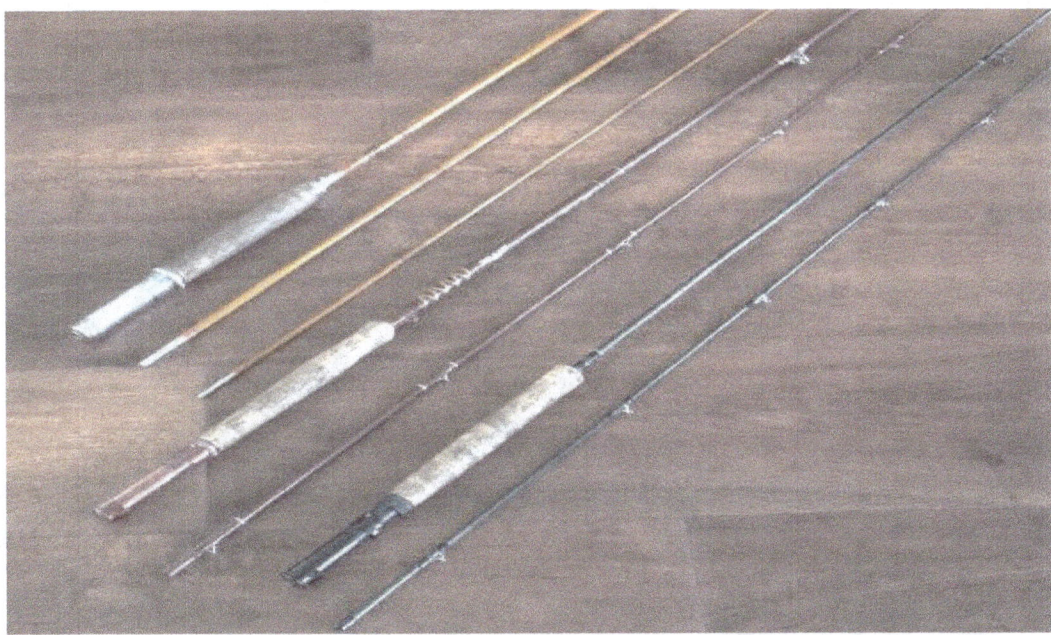

Figure 136 - Fly rods made of various materials. L - R Montague bamboo, Fenwick brown glass, Cortland graphite. Designed and priced at the time for the average angler.

Roll Cast – A method of fly casting where the entire cast is made in front of the angler, by using the water to create drag on the fly line to load the road with enough energy to propel the line forward.

Round Bend – A bend of the hook that is a semicircle. Also known as a Perfect Bend. *Appendix H.*

Running Line – The line to the rear of the head of a fly line designed to be pulled by the heavier head of the fly line. *See also Head. Appendix E.*

S

Sac Fry – *See Alevin*

Saddle Hackle – Long, often webby feathers from where the neck of a chicken becomes the back. *See also Hackle*

Figure 137 – Dyed Saddle Hackle mostly used for streamers.

Salmon Hook – A forged hook with a long shank and either a sproat or Limerick bend. Traditionally these hooks had a black lacquer finish often referred to as a Japanned finish. Hooks for full dress salmon flies are highly specialized and are the subject of much discussion as there are many different designs. *See Sproat Bend, Limerick Bend, Looped Up Eye.*

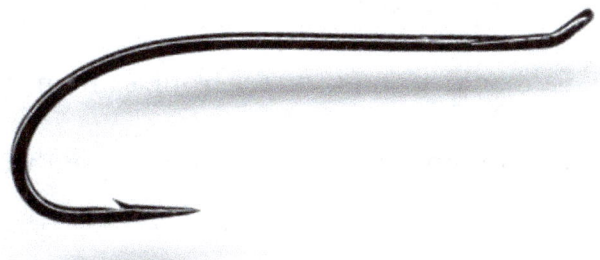

Figure 138 - Salmon fly hook

Salmon Fly or **Full Dress Salmon Fly** – Flies tied on specialty salmon hooks, usually without an eye. The tyer begins by looping a piece of Dacron, monofilament or gut around the shank of the hook to form the eye. The fly is then constructed with as many as nineteen different parts and often using exotic or hard to find materials. Today they are tied more as an art form rather than a fly to be fished with.

Figure 139 Orange Parson full dress salmon fly, Tied and photographed by Tim Trexler.

Salmonidae – A family of ray finned fishes that contain salmon, trout and char and whitefish. Probably the family of fish that is the most sought after by fly anglers. It includes Atlantic Salmon *Salmo salar*, Brown Trout *Salmo trutta*, the char Brook Trout *Salvelinus fontinalis*, Arctic Char *Salvelinus alpinus*, Dolly Varden *Salvelinus malma*, and the Pacific salmon and trout, Rainbow Trout *Oncorhynchus mykiss*, Cutthroat Trout *Oncorhynchus clarki* (along with other species), King Salmon or Chinook *Oncorhynchus tshawytscha*, Silver Salmon or Coho *Oncorhynchus kisutch*, and Sockeye Salmon *Oncorhynchus nerka*. Many other additional species and sub-species of trout and char are found around the world. This family of fish has been and continues to be raised in hatcheries and farms for sport and commercial use. Rainbow Trout and Brown Trout in particular have been distributed around the world to create recreational fisheries, often to the detriment of native species.

Salmoninae – A sub-family of Salmonidae and includes trout, salmon, char, taimens and lenoks.

Salter – A colloquial term referring to brook trout that spend part of their life in brackish or saltwater environments after having been spawned in freshwater.

Saltwater – In this context, saltwater is the ocean, seas, bays, estuaries and inlets that have a high salt content that can make untreated carbon steel hooks very susceptible to corrosion.

San Juan Worms / Squirmy Wormies – Flies that imitate earthworms or large chironomid larvae. Usually tied on a jig style fly hook, and fished beneath a strike indicator. These flies – if you want to call them that - still give some fly anglers a sense of moral superiority over the bait anglers, who are fishing next to them with a live earthworm under a bobber.

Figure 140 - Squirmy Wormies

Scale – (1) A small rigid plate that grows out of the skin of fish. (2) The act of removing the scales of a fish in order to prepare it for consumption.

Figure 141 - The scales and lateral line are visible on this redfish.

Schlappen – Long, webby, and often dull-colored feathers found near the tail of a rooster chicken.

Scientific Method - Make an observation, create a hypothesis, test the hypothesis, form a conclusion based on the observed results of the test.

Scissors – For fly tying the most important tool a fly tyer can own.

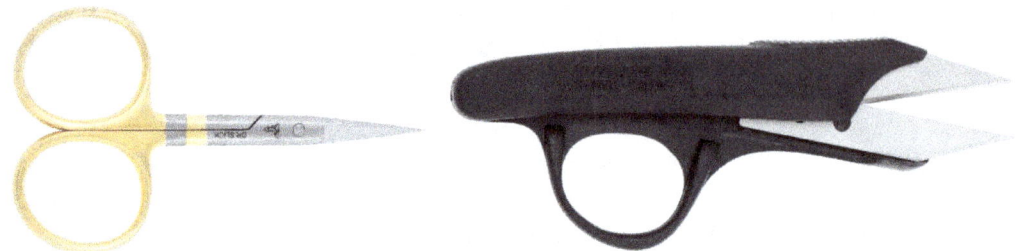

Figure 142 - Two types of fly tying scissors. Photo courtesy of Dr. Slick

Scud Hook – A hook that is between Standard and 2X Short, made of 1X to 2X Heavy wire, with a continuous bend.

Figure 143 -Firehole 321 Scud Hook

Sculpin – A fish that belongs to the superfamily *Cottoidea* whose members may be found in both fresh and marine environments. A bottom dweller, sculpins are a favorite food of trout and are often imitated as streamers. The Muddler Minnow is perhaps the most famous imitation of a sculpin.

Sea Guard – A proprietary finish of the Wright & McGill Company for saltwater hooks.

Seam – A difference in current speeds in a stream often indicated by bubbles, foam or different colored water.

Sediment – Those fine particles of minerals or organic detritus that settle to the bottom of a water body.

Sedimentary Rock – Rocks that are formed on or near the Earth's surface from the compression of sediments. The resultant rocks show the type of deposition from Siltstone to Shale to Sandstone to Conglomerate.

Settleable Solids - Solids in water that left undisturbed will settle to the bottom of a container in a specified time measured in ml/L.

Shank – That portion of a hook between the eye and the bend where the body of the fly is normally constructed. *Appendix H.*

Shock Tippet – Usually used in saltwater fishing, a shock tippet is a section of monofilament or pliable wire that is heavier than the leader to which it is attached. This is to allow for hooking and successful landing of fish with sharp teeth. Also known as a Bite Tippet.

Figure 144 - A Shock tippet made from Rio's knotable Wire Bite

Sighter – A piece of colored monofilament or fluorocarbon built into a leader, for nymph fishing. The movement of the colored leader is a visual cue that a fish has taken the nymph.

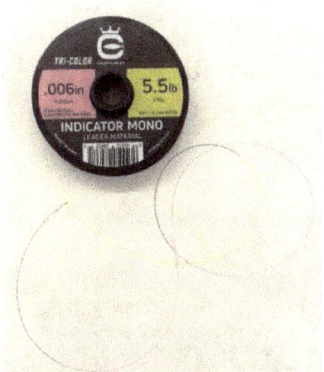

Figure 145 - Tricolor leader material.

Sinker – A weight made of lead, tin, tungsten, some alloy or ceramic that is attached to the fishing line or leader to sink the bait or lure to the desired depth.

Figure 146 - Sinkers come in a variety of sizes and shapes, each with their own use.

Siwash Hook – A standard length hook with an open eye. These hooks are most commonly found on lures, or used as replacements for treble hooks on spinner baits, plugs or other hard lures. Siwash Hooks are sometimes used for tying large, bulky streamers or as Stinger Hooks on articulated streamers.

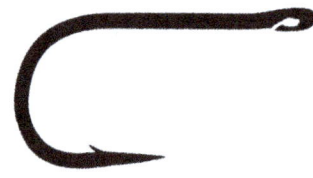

Figure 147 - Siwash hook used for treble replacement

Size also **Hook Size** – A semi-standard rating of the relationship between the gape distance and the length of the hook. *Appendix G.*

Smolt – A young salmon or trout that is beyond the parr stage and ready to migrate to the sea or larger water body.

Snap Swivel – A revolving metal cylinder with a loop at one end to attach to the line and a wire clip at the opposite end to allow for quick removal or replacement of lures. *See also Swivel*

Figure 148 - Snap Swivel

Sneck Bend –A wide gap hook bend with two rounded but distinct corners and the spear is usually upturned. It differs from a Retainer Bend in that the gape is wider than normal and the spear is upturned. *See also Retainer Bend*

Snell – Attaching a short piece of tippet to a hook prior to fishing. It is then said to be a snelled hook. Snelled hooks are usually purchased as ready-made, though anglers often make their own in preparation for a fishing trip. The type of knot used is a snell knot.

Figure 149 - Snelled hooks for bait fishing.

Soft Hackle Fly – A wingless wet fly that is fished under the surface or in the surface film to imitate an emerging insect.

Figure 150 - Soft hackle wet fly. Photo courtesy of Holly Flies.

Soft Plastic – *See Plastic Baits*

Soft Water – Freshwater with a small amount of dissolved calcium or magnesium compounds, generally under 60 mg/L as calcium carbonate. *See also Hard Water*

Spade – For hooks without eyes, the front portion of the hook is flattened with the wider section being near the front end of the hook. The "tippet" or "snell" is then lashed to the hook and the wide part of the hook prevents the lashings from slipping.

Spawn – The act of reproduction for fish. *See also Redd.*

Spear – That portion of the hook from the heel of the bend to the point of the hook. *Appendix H.*

Species – A taxonomic unit below genus capable of normal breeding. Denoted along with genus as in lower case e.g. *Salmo salar* where *salar* denotes the species.

Spectrum, electromagnetic – Radiation from long wavelength radio waves to short wavelength gamma rays. The visible light spectrum is between 400 and 790 terahertz (THz) or the wavelengths are roughly 390 – 750 nanometers (1 nanometer = 1 billionth of a meter. The electromagnetic spectrum is not only responsible for visible light, but for microwaves, radio waves, x-rays and gamma rays which have multiple uses in the modern world for everything from deep space imaging, to fighting cancer to cooking your breakfast burrito.

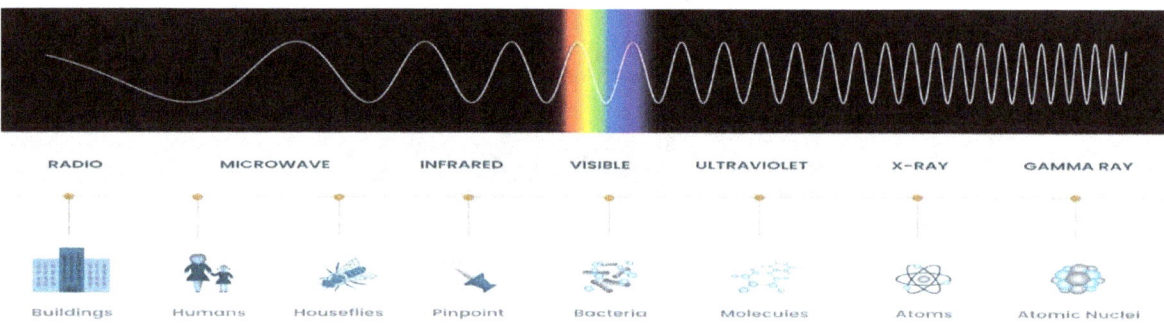

Figure 151 - Electromagnetic spectrum. Photo courtesy NASA

Spey Fishing – A fishing technique originally developed in Scotland using long rods, making long casts with two hands on the rod to catch salmon and steelhead.

Spey Fly – A fly tied on a light salmon fly hook, with a body that does not occupy the entire hook shank, a sparse wing, and long soft hackle that extends to the bend of the hook and beyond. The flies are used primarily for taking salmon, sea-run steelhead and brown trout.

Spin Casting Reel – *See Closed Face Spinning Reel*

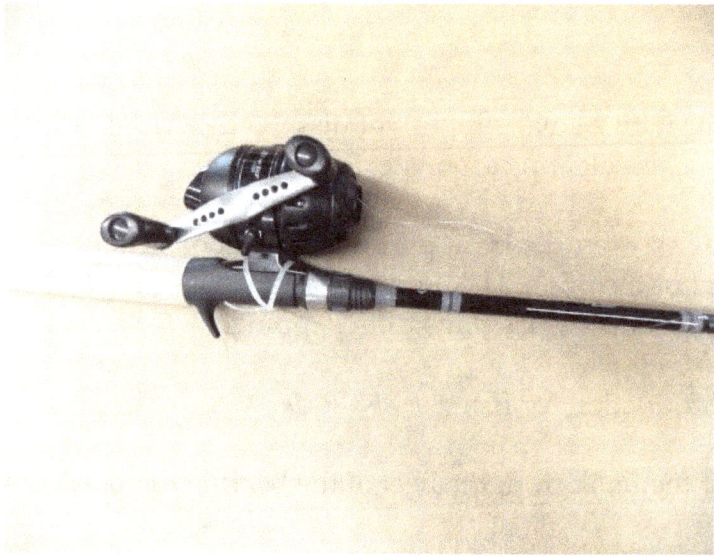

Figure 152 - Spin Casting Rod and Reel. A good choice for beginners.

Spin Fishing - Casting a lure from a spinning reel that is then retrieved to imitate bait. In Spin Fishing the weight of the lure and the action of the rod determine the distance of the cast. *See also Fly Fishing, Open Face Spinning Reel.*

Figure 153 - Spin fishing

Spinner – (1) An adult mayfly that is mating or has mated and fallen to the water. They usually possess large eyes, clear wings, bright colors and no mouthparts. Flies tied to imitate them usually lack hackle. (2) A fishing lure that has a wire shaft, a blade that revolves around the shaft, a body of plastic or metal and a hook at the back end. They are not usually fished with fly tackle.

Figure 154 - Typical spinner lure.

Spinner Bait – A V–shaped wire lure with one or two blades on the top of the V and a hook with a weight molded to it on the bottom that is sometimes dressed with a fringe rubber skirt. Usually fished with a baitcasting outfit or a spinning outfit.

Figure 155 - Spinner Baits

Spinner Fall - Following mating the female mayflies deposit their eggs on the water, the spent insects, having completed their primary purpose in life fall to the water to die. This fall of insects can induce major surface feeding by fish.

Split Ring – A small wire ring of three loops that is used to attach hooks to lures in hardware fishing.

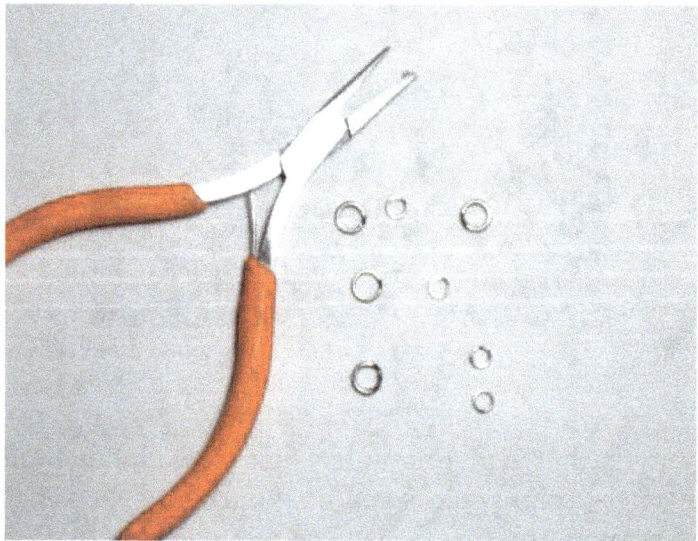

Figure 156 - Split rings with split ring pliers. The small hook on the plier's jaw facilitates easy use of the rings.

Split Shot – Round weights made of lead, tin or some alloy that have a groove cut halfway through to allow them to be attached to a leader by compressing the groove.

Figure 157 - Split shot is pressed onto the leader. Pushing the ears will open it for removal.

Spiracle –One of a series of small holes used for breathing in arthropods on land.

Spring – The point where the ground surface and aquifer meet and water flows out. *Appendix B*

Spoon – A metallic lure, usually oblong in shape with a hole at one end to attach to the line and a hook fastened to the opposite end. The original incarnations of this lure were made from dinnerware of the same name.

Figure 158 - Daredevil spoons.

Sport Fishing – The act of fishing because of the enjoyment derived from the activity, not because of necessity for sustenance or monetary gain. *See also Sustenance Fishing and Commercial Fishing.*

Sproat Bend – A hook bend that is a parabolic (as opposed to a round bend which is semicircular) that decreases in radius as it nears the heel of the hook. The name is largely believed to come from W. H. Sproat, a 19th century Englishman who devised this particular hook bend.

Squirmy Wormy – *See also San Juan Worm and Plastic Bait*

Stacker or **Hair Stacker** – A fly tying tool used for evening the tips of the hair to tie flies.

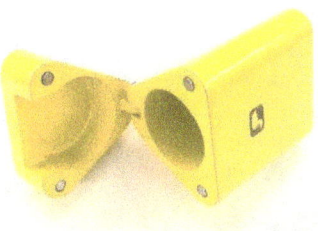

Figure 159 - Hair Stacker

Stainless Steel – A type of steel that contains at least 10.5% chromium and less than 1.2% carbon. It is corrosion resistant and is thus useful for making saltwater fishing hooks.

Standard – In hook making an ambiguous term, determined by each manufacturer as to wire diameter, hook size, hook length and any other feature they wish to attribute to a hook. In flylines measurements based on weight of a length of line by the AFFTMA. *Appendix G, Appendix E.*

Steelhead – Rainbow trout that move out of their natal waters and live most of their lives in a larger water body, before returning to their natal streams to spawn. Those that migrate to the ocean and return to freshwater are referred to as anadromous. Those that move to large freshwater lakes, such as the Great Lakes are referred to as potamodromous.

Stickbait or Stick Bait – A hard-body lure that usually resembles baitfish, or some other aquatic organism. Sometimes called a jerk bait, or by an older term of plug, and most often fished with a baitcasting outfit or a spinning outfit.

Figure 160 - Stickbaits

Stinger Hook – (1) Usually in the angling world, a stinger hook is a hook that is attached by flexible connection behind the main lure. In fly fishing this is the case as well, such as in two hook, articulated streamers. (2) A large gape, long shank hook on which hairbugs or other large streamers are tied.

Figure 161 - Stinger hook for tying spun hair bugs.

Stochastic – That which involves a random variable.

Stocked Fish – A fish that had been spawned and raised in an artificial environment and then transplanted to a body of water. *See also Wild Fish and Native Fish.*

Stocking – The artificial introduction or re-population of organisms into a habitat. Often game animals or fish intended to be harvested.

Stonefly – An insect of the order *Plecoptera*. The nymph living in water resembles the adult without wings. They generally have long antennae, weak chewing mouthparts and thoracic gills. The adults have two pair of wings that lay flat over the back and they are not strong fliers. *Appendix J.*

Stream Order – A first order stream has no tributaries. Two first order streams combine to form a second order stream, and so on. *Appendix C.*

Streamer – A longer fly usually tied imitate baitfish, but sometimes tied to simply attract fish. This lure is usually constructed with feathers, though natural and synthetic hair may also be incorporated into the lure. Streamers are fished underwater, by actively stripping it in, drifting it with the current or trolling it behind a boat. *See also Bucktail*

Figure 162 - Black Ghost Streamer. Photo courtesy of Holly Flies

Strike Indicator – A buoyant device attached to a leader used in nymph fishing to indicate when a fish has taken the fly. *See also Bobber*

Figure 163 - Three different types of strike indicators.

Stringer – A piece of rope or a chain with large snaps to keep fish.

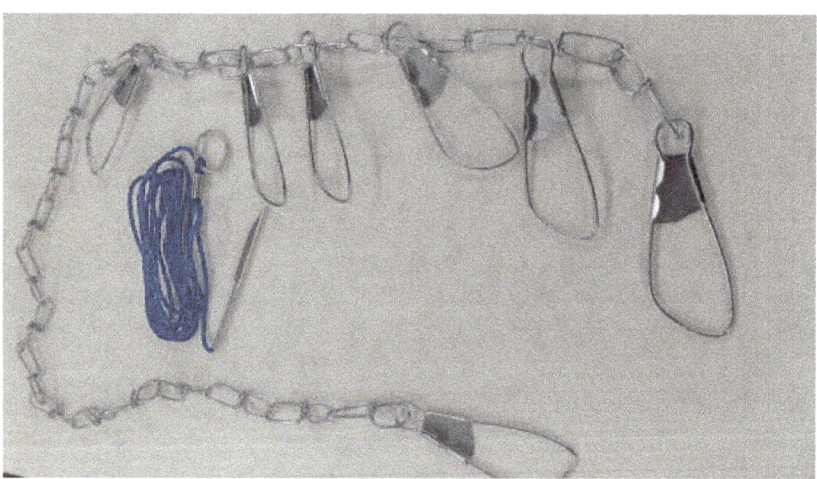

Figure 164 - String and chain stringers.

Sub-Surface or Subsurface – That portion of the water column that is below the surface of the water.

Succession – In ecology, an ecosystem which began as a simple system with few species of the same age moving toward a system with many different species of varying ages.

Sucker Spawn – A fly tied to resemble the eggs of suckers following spawning.

Figure 165 - Sucker Spawn. Photo courtesy of Holly Flies.

Sunfish – Those freshwater fish of the family *Centrarchidae* consisting of eight genera. These include the Black Bass (*Micropterus*), the true sunfishes (*Lepomis*), crappie (*Pomoxis*), rock bass (*Ambloplites*), banded sunfishes (*Ennecanthus*), Sacramento perch (*Archolites*), mud sunfish (*Acantharchus*) and flier (*Centrarchus*). Except for the Sacramento perch, members of the sunfish family were largely confined to North America east of the Rocky Mountains. Widespread stocking has scattered various genera around the world. Members of the genus *Lepomis* readily interbreed, producing offspring with morphological features of both parents. Many anglers began their angling adventures catching sunfish, particularly bluegills (*Lepomis machrochirus*), redbreast sunfish (*Lepomis auritus*), green sunfish (*Lepomis cyanellus*), or pumpkinseed (*Lepomis gibbosus*). The fish are popular quarry no matter the choice of tackle and have a reputation as fine table fare.

Figure 166 - Bluegill taken through the ice.

Surface Film or **Surface Tension** or **Surface Film Tension** – The property of the surface of a liquid that is a stretched elastic membrane. This tension is what allows insects and dry flies to walk on the water surface. They do not possess enough weight to break the surface tension.

Suspended Solids - Particles in water that are larger than 2 microns and will not settle out within a measured time (30 minutes in many laboratory test situations). To measure them a given volume of water is passed through a previously weighed 2 micron filter and the filter is dried and weighed again. Measured in mg/L

Sustenance Fishing – Fishing to collect fish because it is a basic part of one's diet. *See also Sport Fishing and Commercial Fishing.*

Swaging – The process of pushing metal into a die to form it to a specific shape.

Swamp – An area of low lying ground with water present that is predominantly forested. *See also Marsh.*

Figure 167 - Swamps have trees, marshes are herbaceous.

Swim Up Fry – Recently hatched fish that have used up the egg sac for food and swim to the surface to gulp air to fill their swim bladder and can then begin to feed.

Swivel – A small metal cylinder with revolving loops at each end to which the line and leader are fastened. The purpose of the swivel is to prevent line twist caused by the action of the lure or bait. *See Snap Swivel.*

Figure 168 - Barrel swivel and 3 way swivel

Sympatric – Two related species or populations of either plants or animals that exist in the same area.

Synergy – The interaction of two organizations, individuals or substances that acting together produces a result that is greater than either of the other two individually.

T

Taxa – A unit of classification of an organism. *See also Taxonomy*

Taxonomic Rank – The place assigned for each organism from broad at the top to definitive at the bottom.

> Domain
> Kingdom
> Phylum (in zoology) / Division (in botany)
> Class
> Order
> Family
> Genus
> Species

Taxonomy - The science of the systematic study of the classification of organisms.

Tempering – Heat treating steel is usually a two-step process. It is first hardened (most often referred to as quenching) followed by tempering. Steel is hardened by heating it to a temperature range where its molecular structure changes (red hot ~1600 degrees F) and then rapidly cooling it by quenching in water or oil. This results in high hardness but low ductility. Heat treating changes the hardness (strength), ductility (ability to deform without fracturing), and toughness (ability to withstand impact without fracturing) to the desired properties. For steel of a given composition, as hardness is increased, ductility and toughness decrease. It is important to get this right when making wire for fly hooks. *See also Anneal*

Tenkara – A form of fly fishing popularized in Japan that involves the use of a long collapsible rod without a reel, where the line is attached to the tip of the rod. The angler drops or daps the fly on the water rather than actually casting. Fighting the hooked fish is dependent on the suppleness of the rod. *See also Dapping.*

Figure 169 - Tenkara Fly. Photo courtesy Mary Kuss.

Tenkara Fly – A style of fly developed in Japan. Almost any fly can be fished with a Tenkara rod but, Tenkara flies are simply tied wet flies, similar to soft hackles. One large difference is in their original design Tenkara flies do not match specific hatches. Some Tenkara flies are tied with the hackle facing forward over the eye of the hook. These reverse hackle flies are known as Sakasa Kebari (which means "backwards fly" in Japanese). The backward hackle will act as a drogue chute helping keep the line tight.

Tensile Strength – The maximum load that a material can support without fracturing that is measured in units of force per unit area. The units are pounds per square inch (psi) in the English system of measurement or grams per square centimeter (g/cm^2) in the metric system.

Terrestrial – (1) A land-based insect. (2) A fly tied to imitate those land-based insects that are important as fish food. They include but are not limited to grasshoppers, crickets, ants, beetles, inchworms, cicadas, bees, leafhoppers including jassids, and so on. Terrestrial flies were popularized in the 1960s and 1970s in the Cumberland Valley in Pennsylvania by Vince Marinaro, Ed Shenk and Charlie Fox.

Figure 170 - Figure 96 - Terrestrials Clockwise from top left, Ant, Beetle, Shenk's Letort Cricket, Letort Hopper. Photo Courtesy of Holly Flies

Test or Pound Test – The number of pounds (or kilograms) of weight in a standard test that will break a line or leader. The line is then rated as XX lb.

Figure 171 - Spools of heavy leader materiel with pound test rating on left side, diameter in the center.

Thalweg- In a stream, the lowest point or the deepest sections of the stream that is the main line of flow.

Thermocline – The transition layer between warm water at the surface and cold water at the bottom of a water body. *Appendix B.*

Thermodynamics, First Law - Also known as The Law of Conservation of Energy. Energy is neither created nor destroyed by any physical or chemical process. It is merely transformed from one form to another.

Thermodynamics, Second Law - Any system plus its surroundings tends spontaneously toward increasing disorder or randomness.

Thorax Fly – A modification the traditional Catskill Style dry fly. In a Thorax Fly the wings are set back farther from the eye at about the middle of the shank and the hackle is wrapped on both sides of the wing, and trimmed to allow the fly to float closer to the surface of the water.

Figure 172 - Thorax fly. Photo courtesy of Holly Flies

Thread – The fundamental material needed to tie flies. A thin string material made from silk, cotton or synthetic materials that is wound around the hook to act as part of the fly body, or attach materials. Measured as ought where 6/0 is finer than 3/0, or by Denier where 70d is finer than 100d. *See also Denier*

Figure 173 - Tying thread comes in multiple sizes and colors

Throat – On a hook, the horizontal distance from a perpendicular line between the point and the shank directly to the farthest part of the inside of the bend. *Appendix H.*

Tippet – The front end of a leader that is attached to the hook. This is usually made of nylon or fluorocarbon and sometimes wire, and is the final section of line between the angler and the fish. *See also Leader, Dropper, Appendix F.*

Figure 174 - Tippet

Tippet Ring – A steel ring that comes is sizes from 1.5 to 4.5 mm and is tied to the leader and to which the tippet is attached on the opposite side.

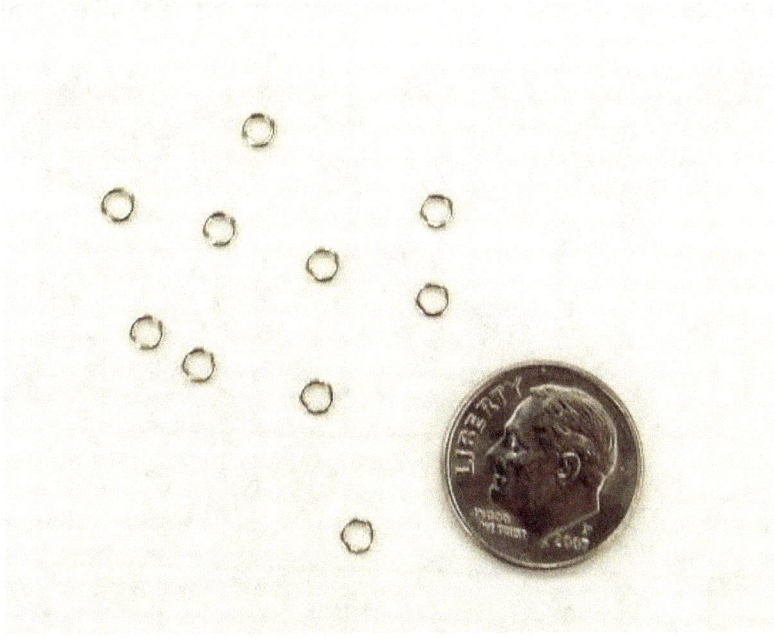

Figure 175 - 2 mm tippet rings

Tip Up – A device used in ice fishing which is set over a hole in the ice where bait is on a line beneath the ice in the water, and a flag or other indicator will release and spring up when the bait is taken.

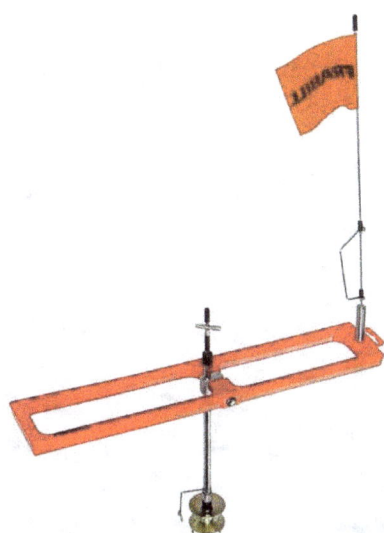

Figure 176 – Tip Up used in ice fishing. Line is wound around the lower spool. A hook and leader is at the end. The whole mechanism is placed over the hole and the flag is bent to a down position and springs up when a fish takes the bait and the spool begins to revolve. Photo courtesy of The Tackle Shack, Wellsboro, PA.

Topographic – A description of the physical features in a given area and how they spatially relate to each other. A topographic map shows land contours, water courses, and other physical features, as well as some cadastral and social features such as road, stream and township names.

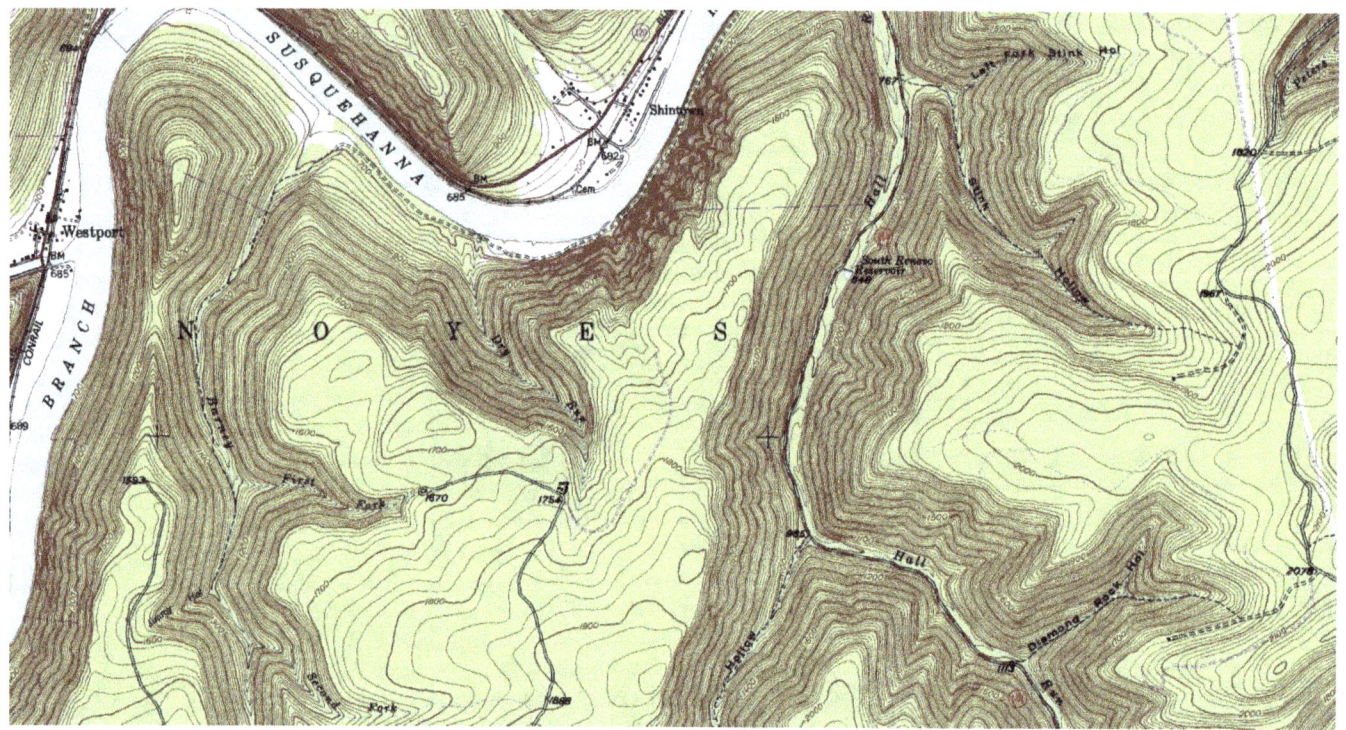

Figure 177 - Topographic maps show elevations by tracing contours in brown lines. Contours that meet themselves are the high points. Streams flow downhill denoted by inverted Vs. The bottom of the V points upstream.

Trailer Hook – A hook that is attached behind the fly, or the second and third flies in articulated patterns. *See also Articulated Fly, Intruder Fly.*

Travel Rod – A fishing rod with multiple sections that can be easily packed in luggage for travel.

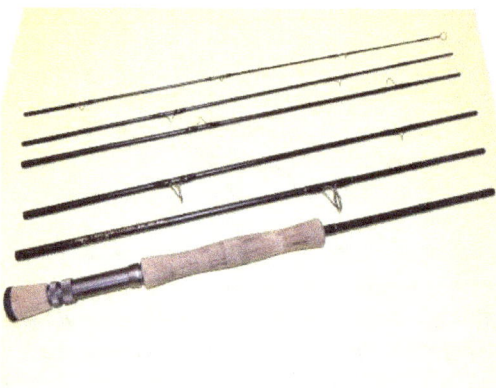

Figure 178 - 6 piece travel rod

Treble Hook – A hook with three points that has been welded, brazed, or soldered together at the shanks. Some fly companies produce a salmon hook for fly tying, but most treble hooks are used for hardware applications.

Figure 179 - Treble hook

Trolling – A method of fishing where a lure, bait or sometimes multiple flies or streamers are pulled along behind a boat or canoe that is moving.

Figure 180 - Trolling for walleye on Lake Erie

Trophic - That which relates to feeding or nutrition. Levels are defined by food categories. For example a fish whose primary diet is other fish functions at a different trophic level than a fish whose primary diet is insects.

Trout – Members of the family *Salmonidae*. They are characterized by having dark spots on a light background. The most common trout found by anglers in America are the Brown Trout (*Salmo trutta*), the Rainbow Trout (*Oncorhynchus mykiss*), and the Cutthroat Trout (*Oncorhynchus clarkia*). Many anglers when discussing trout also lump in the misnamed Brook Trout (*Salvelinus fontinalis*) which is actually a char, a genus distinct from the two genera of trout in North America. Trout are a prized gamefish and fine table fare. They are sought by millions of anglers who spend billions of dollars on the sport. Because of their fine culinary appeal they are also raised commercially for the table. Because of over-exploitation and habitat loss stocking is used in some streams and lakes to maintain populations for anglers to fish for. Most stocking is Put and Take, in that the fish are expected to be caught and kept by the time the season ends. *See also Char, Salmonidae.*

Tube Fly – A type of fly that originated in 1945 and is credited to Minnie Morawski, of Aberdeen, Scotland. The flies are tied on thin tubes, rather than hooks. The tubes are made of brass, copper, aluminum, steel, or plastic. The tippet is threaded through the tube and attached to a "trailer hook" which is usually a short shanked, wide gap hook. The tube fly will be ahead of the mouth - and thus the teeth – of the fish, preventing major damage to the fly. Mostly tube flies are tied for salmon and steelhead.

Turbidity - The optical measurement of relative water clarity.

U

Umwelt - The world as it is sensed and perceived by an individual organism. Organisms can share the same environment but have much different umwelten (plural).

Up Eye – The eye of the hook that is bent upward from the shank, usually about a 45° angle, but may vary between 0° and 90°.

Figure 181 – Midge Fly hook with up eye

V

Vane – Used for stream habitat repair they are sometimes called rock vanes or log vanes, they are structures placed in a stream with the downstream end anchored to the bank and the upstream end closer to the center of the stream. The device will allow water to flow over it and turn into the center of the stream, preventing bank erosion and keeping the water in the center of the stream.

Figure 182 - Two log vanes armored with rocks to keep a narrow stream in its channel.

Vertical Plane – An imaginary plane that extends from the end of the eye of the hook to the farthest reach of the back of the bend and is parallel to the spear and point of the hook. The vertical plane is important to streamer fly patterns as well as many salmon fly patterns, because a fly tied in a straight vertical plane allows it to track straight through the water.

Vise – A mechanical device used to hold hooks for fly tying. Often incorrectly misapplied and misspelled as vice which includes things like drinking, smoking and playing cards and/or presidential sidekicks.

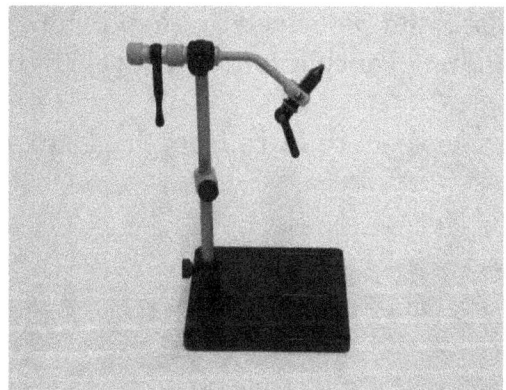

Figure 183 - Fly tying vise

In Vitro and **In Vivo** – In Vitro is Latin for "within the glass," meaning some biological process happening artificially. In Vivo is Latin for "within the living," meaning in nature. Fish eggs in a hatchery may be said to be fertilized and hatched In Vitro, while spawning wild fish fertilize their eggs and they hatch In Vivo.

W

Waddington Shank – A double wire shank for tying flies where an undressed hook (mostly small treble hooks) are looped onto the shank through a break in one of the wires.

Waders or Chest Waders - Chest high waterproof boots, which come with the boot attached (Bootfoot waders) or a flexible fabric foot (Stockingfoot waders) with which wading shoes must be worn.

Figure 184 - Stocking foot chest waders. Photo courtesy of Frogg Toggs

Wading Shoes – Shoes specially designed to be worn over the foot of stockingfoot waders, with various types of soles to accommodate and provide traction for the wearer on the intended stream substrates.

Figure 185 - Wading shoes with studded soles and Bowa lace.

Wading Staff – A sturdy pole of various materials that helps the angler keep his balance.

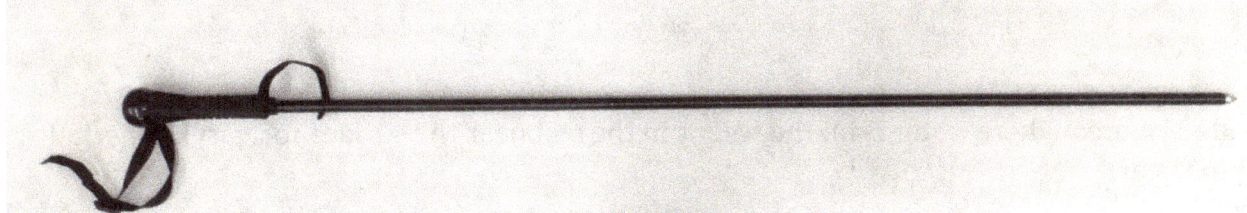

Figure 186 – Collapsible Wading Staff

Waist Highs – Waders that come up only to the person's waist. May be either bootfoot or stocking foot.

Figure 187 - Waist high stockingfoot waders. Photo courtesy of Frogg Toggs

Watershed – The area where the water that falls on it flows downstream to a common point. *Appendix C*

Warmwater – In this context, this is meant to mean those freshwater systems – rivers, lakes and ponds – that contain fish species more tolerant of water temperatures over 60ºF. Warmer water holds less oxygen. These fish may include but not be limited to the black basses, the pike family, the walleye family, panfish, catfish, carp and other fish.

Wet Fly – In this context, those flies tied to imitate aquatic insects that are maturing and moving about underneath the water surface in preparation to maturation. These include traditional wet flies, soft hackles, and emergers. *See also Emerger*

Figure 188 - Gold Ribbed Hares Ear wet fly. Photo courtesy of Holly Flies

Wetland – An area where water is on the soil or in the soil near the surface for varying periods during the year.

Whip Finisher – A tool used in fly tying to finish the head of the fly by wrapping the thread underneath itself.

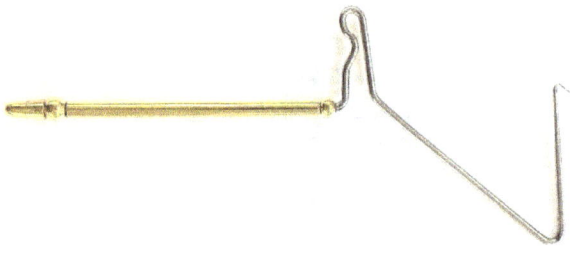

Figure 189 - Whip finisher tool.

Wild Caught – Fish raised in an environment that are captured in a sustainable way for food.

Wild Fish – A fish that has been born in the water where it lives. The original stock of fish – brown trout for example – may not have been originally found in the water body but the stocked fish successfully spawned there. *See also Native Fish and Stocked Fish.*

Figure 190 - A wild stream bred brown trout, not native to North America.

Wire – A thin metal cylinder that is the basic material for hook construction.

Wooly Bugger – A streamer with a marabou tail, chenille body and saddle hackle palmered over the body. The invention of this fly is credited to the late Russ Blessing of Harrisburg, PA.

Figure 191 - Wooly Bugger. Photo courtesy of Holly Flies

Worm Hook – A hook with two right angle bends immediately behind the eye. The hook was originally developed for fishing plastic worms where the hook point is embedded in the worm and the eye sticks out the front of the worm. In recent years this type of hook has gained some stature in the fly tying community as a hook for large bulky streamers that are resistant to hanging up on weeds and other underwater debris.

Figure 192 - Worm hook

X

X Factors – (1) A method of notating the relationship to a standard in fishing hooks.

X Fine or 1X Fine denotes wire diameter 1 size smaller than a standard wire for that size hook.
X Heavy or 1X Heavy denotes wire diameter 1 size larger than a standard for that size hook.
X Long or 1 X Long denotes a shank length (usually – sometimes it is total length) 1 size longer than a standard length for that model of hook.
X Short or 1 X short denotes a shank length (sometimes it is Total Length or Manufacturer Length) 1 size shorter than a standard length for that model of hook.

This is manufacturer dependent.

Usually 2X Short and 3X Fine are the lower limits for length and weight before the hook becomes ineffective. *Appendix G.*

(2) A system for rating leader and tippet diameter. The hook wire and length rating is not the same for leaders and lines with X nomenclature, where X – such as 2X - is actually referring to a size and where each X rating higher is smaller than the previous one. 2X > 5X > 7X. *Appendix F.*

Y

York Bend – A modified sproat Bend that begins on the shank of the hook between the eye and above the end of the point and continues through the spear of the hook, making the point rise above the bottom of the spear. This is an intermediate between a continuous bend and a sproat bend. A York bend is found mainly on salmon hooks.

Z

Zebra Midge – A small fly tied to imitate the pupa of small aquatic chironomids. It is usually tied on a curved hook in sizes 18 – 26, with a bead head, thread body and wire rib.

Figure 193 - Zebra Midge. Photo courtesy of Holly Flies

Zonker – Streamers with strips of fur, usually rabbit, attached to the shank of the hook.

Figure 194 - Zonker. Photo courtesy of Holly Flies.

Zoology – The study of animals.

Zooplankton – Microscopic aquatic invertebrate animals. Microinvertebrates

Appendices

Appendix A – A Sense of Where You Are

Appendix B - Lake Terminology and the Water Cycle

Appendix C – Watersheds

Appendix D - Fish Terminology

Appendix E - Basic Fly Casting

Appendix F - Average Tippet Diameters and Break Strengths

Appendix G - Average Hook Sizes

Appendix H - Hook Terminology

Appendix I - Fly Proportions and Nomenclature

Appendix J – Insects

Appendix K – Some Thoughts on Hook Sizes

Appendix L – Life List of Fish

Appendix A
A Sense of Where You Are

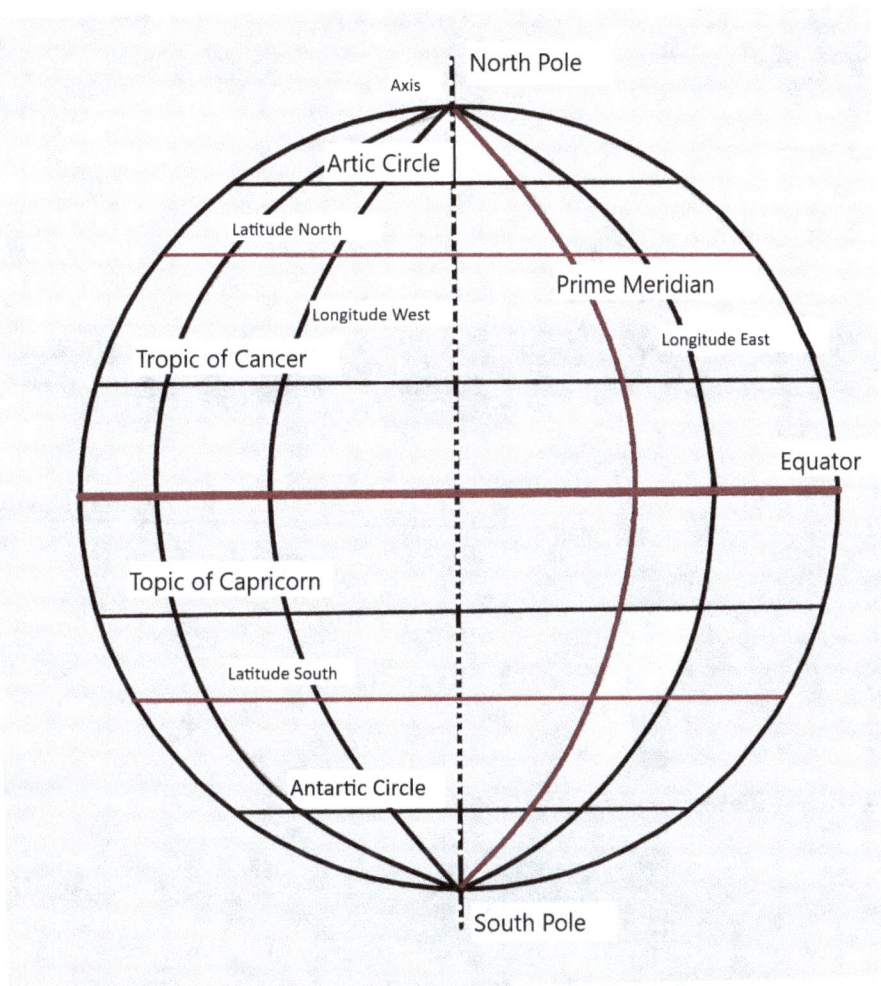

The Equator is zero degrees 0° latitude and runs horizontally around the Earth

Latitude is parallel lines that run around the Earth parallel to the Equator

Longitude are lines that run from pole to pole and converge at the poles.

The Prime Meridian is 0° longitude and runs from pole to pole through Greenwich, England

The Axis runs from pole to pole through the center of the Earth and the Earth spins on it. The Earth is tilted 23.5 degrees on its axis. This is why there are seasons.

When the sun crosses the Tropic of Cancer summer begins in the Northern Hemisphere

When the sun crosses the Tropic of Capricorn summer begins in the Southern Hemisphere

Latitude and Longitude are reported in degrees, minutes and seconds such as

40°15'32"N (latitude) 78°10'07"West (longitude)

Appendix B
Lake Terminology and the Water Cycle

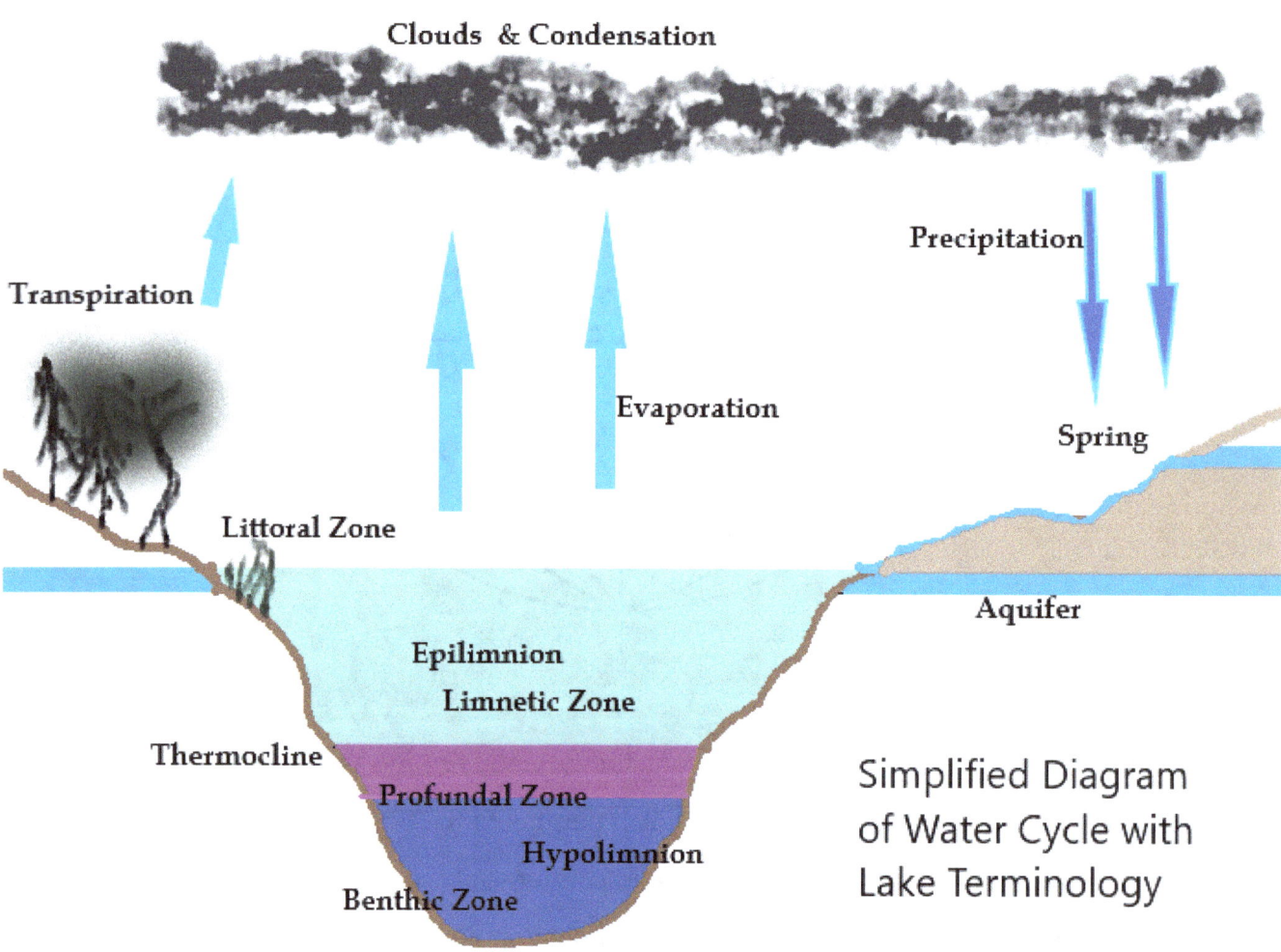

Clouds & Condensation

Transpiration

Precipitation

Evaporation

Spring

Littoral Zone

Aquifer

Epilimnion

Limnetic Zone

Thermocline

Profundal Zone

Hypolimnion

Benthic Zone

Simplified Diagram of Water Cycle with Lake Terminology

Appendix C – Watersheds

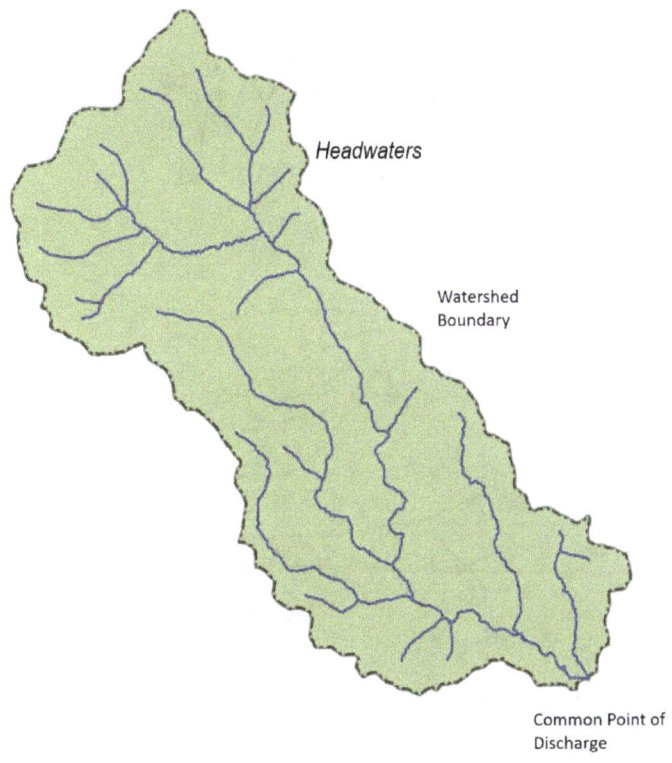

Watersheds may be as large as half a continent such as the Mississippi River or the Amazon River or they may be as small as the drainage area of a first order unnamed tributary. Two first order streams form a second order stream and so on.

The watershed boundary is the topographic high point. Water that falls on the outside of the boundary flows to a different common point of discharge.

Appendix D
Fish Terminology

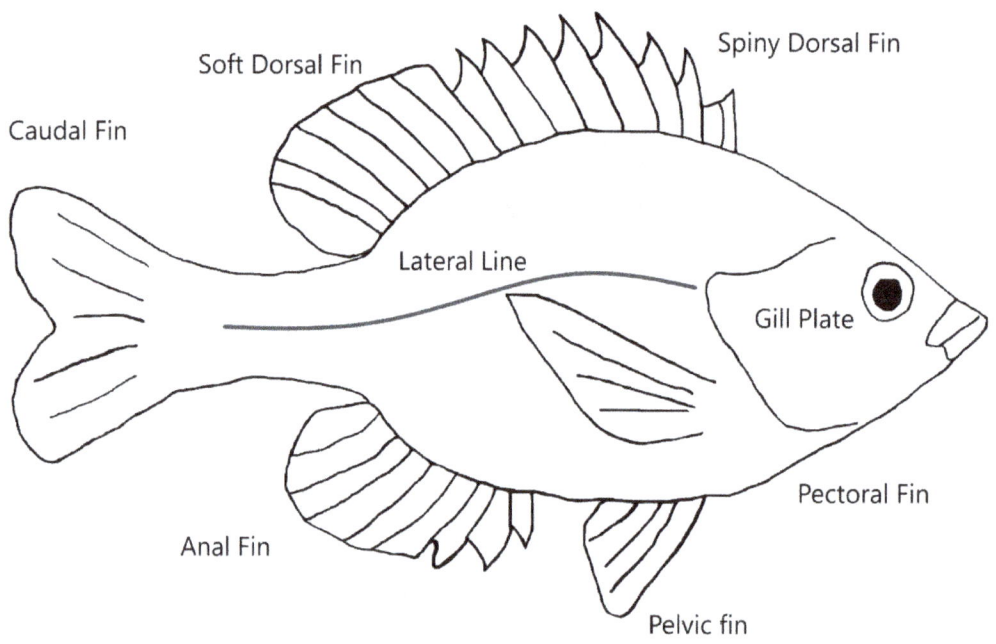

Soft Dorsal Fin

Spiny Dorsal Fin

Caudal Fin

Lateral Line

Gill Plate

Pectoral Fin

Anal Fin

Pelvic fin

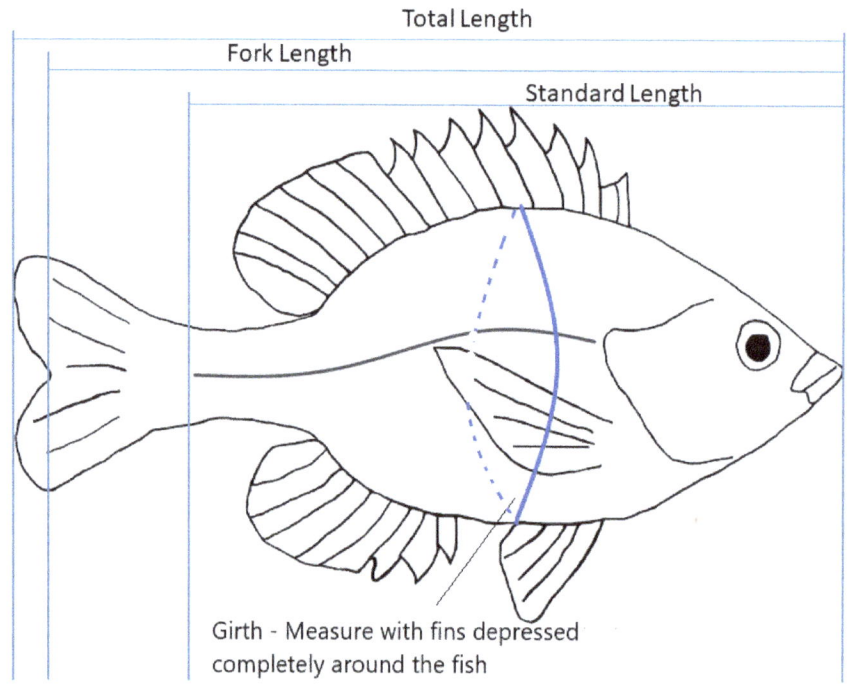

Total Length

Fork Length

Standard Length

Girth - Measure with fins depressed completely around the fish

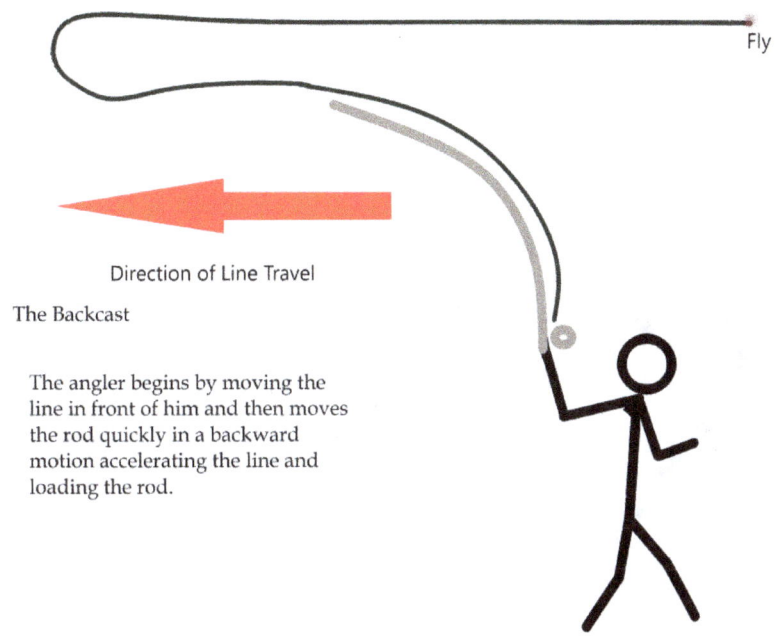

Direction of Line Travel

The Backcast

The angler begins by moving the line in front of him and then moves the rod quickly in a backward motion accelerating the line and loading the rod.

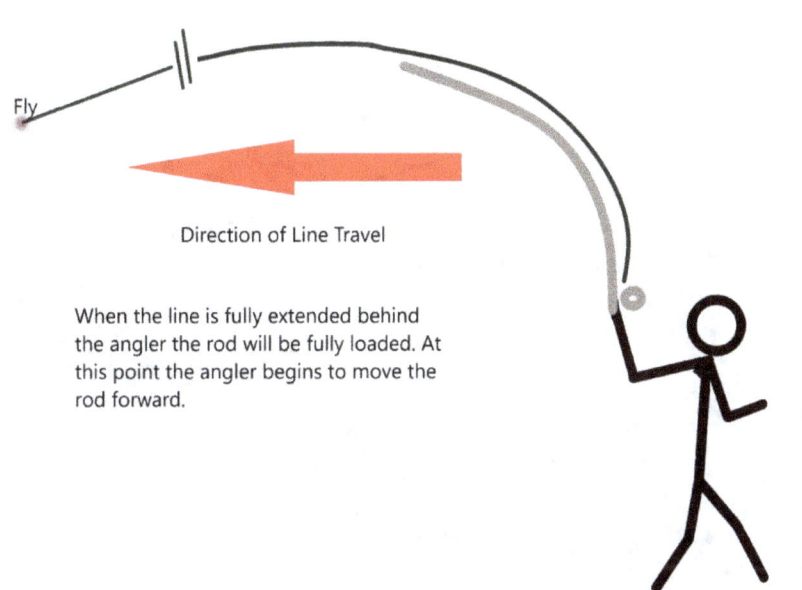

Direction of Line Travel

When the line is fully extended behind the angler the rod will be fully loaded. At this point the angler begins to move the rod forward.

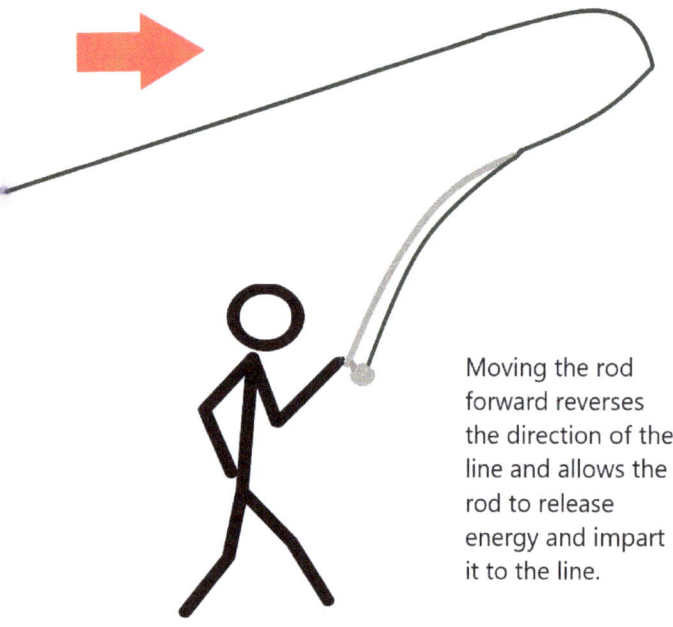

Moving the rod forward reverses the direction of the line and allows the rod to release energy and impart it to the line.

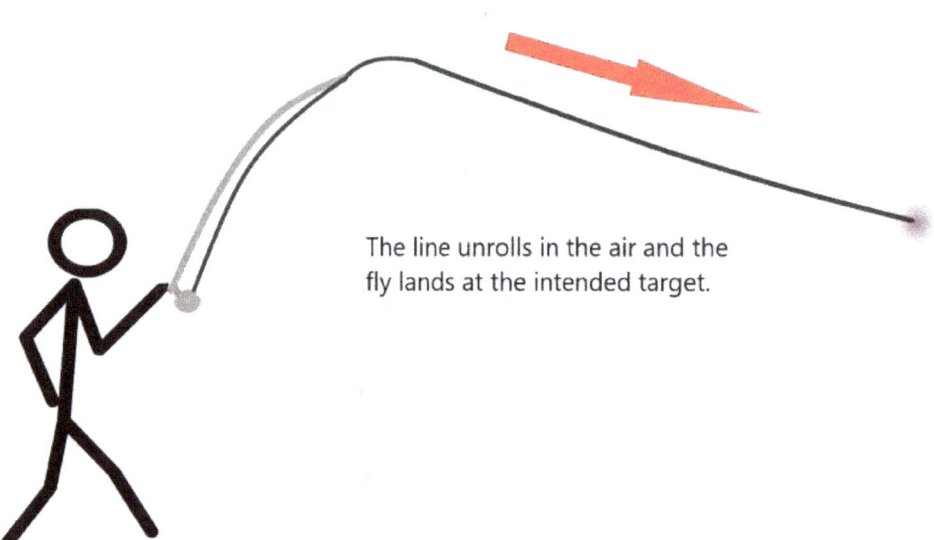

The line unrolls in the air and the fly lands at the intended target.

Of course there is more to fly casting than these simplified diagrams and practice is required to become competent in the technique.

Below is a basic diagram or a total fly line system. Lengths of backing, running line, head, leader and tippet will vary widely depending on the circumstances.

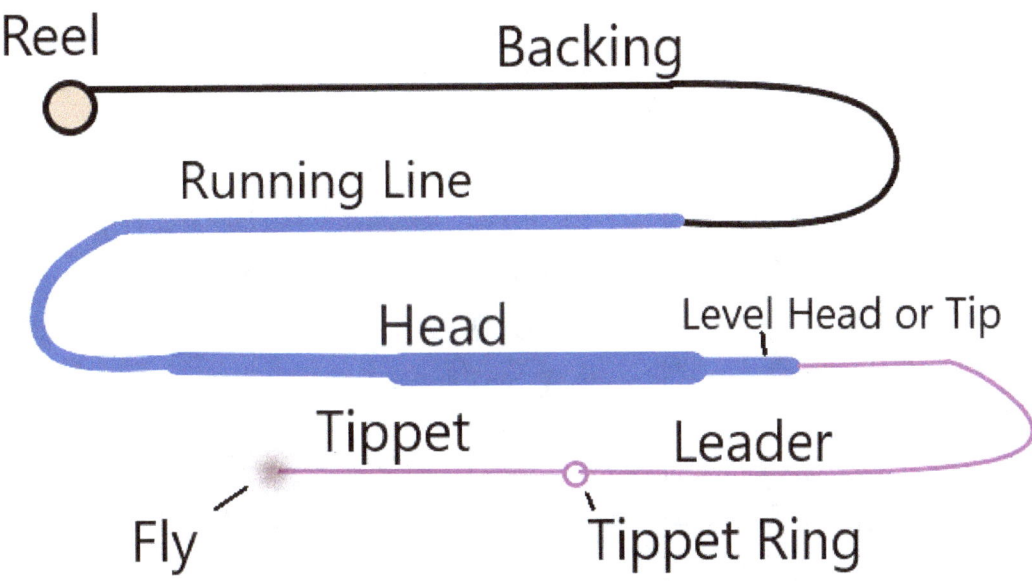

AFFTMA Fly Line Rating

Weight is based on the first 30 feet of line minus the level tip in Grains		
Line Rating	Weight Range Grains	Median Target Grains
1	54 - 66	60
2	74 -86	80
3	94 - 106	100
4	114 -126	120
5	134 - 146	140
6	152 - 168	160
7	177 - 193	185
8	202 - 218	210
9	230 - 250	240
10	270 - 290	280

Appendix F
Average Tippet Diameters and Lbs. Test Break Strengths

Monofilament Tippet		
X Size	Diameter Inches	Lb. Test
0X	0.011	16.90
1X	0.010	14.25
2X	0.009	11.45
3X	0.008	9.00
4X	0.007	7.40
5X	0.006	5.50
6X	0.005	3.80
7X	0.004	2.60

Fluorocarbon Tippet		
X Size	Diameter Inches	Lb. Test
0X	0.011	15.70
1X	0.010	13.10
2X	0.009	11.05
3X	0.008	9.20
4X	0.007	7.85
5X	0.006	5.45
6X	0.005	3.85
7X	0.004	2.75

Appendix G
Average Hook Sizes
Based on Measurements by the Author

Hook Size	Average Gape (mm)	Average Total Length (mm)
1	12.2	35.7
2	10.8	32.2
4	8.9	28.0
6	8.1	23.5
8	7.2	19.9
10	6.0	16.9
12	5.2	14.5
14	4.5	12.5
16	3.9	10.4
18	3.4	8.9
20	3.1	7.8
22	2.7	7.1
24	2.4	6.6

Hook Size	Bead Size (mm)	Bead Size Fraction Inch
6 - 8	4.6	3/16
8 - 10	4.0	5/32
8 - 12	3.8	5/32
10 -12	3.3	1/8
10 -14	3.0	1/8
12 - 14	2.8	7/64
16 - 18	2.4	3/32
16 - 20	2.3	3/32
18 - 20	2.0	5/64
20 - 24	1.5	1/16

Measurements

Millimeters to Inches				
mm	decimal in		mm	decimal in
1	0.04		28	1.10
2	0.08		29	1.14
3	0.12		30	1.18
4	0.16		31	1.22
5	0.20		32	1.26
6	0.24		33	1.30
7	0.28		34	1.34
8	0.31		35	1.38
9	0.35		36	1.42
10	0.39		37	1.46
11	0.43		38	1.50
12	0.47		39	1.54
13	0.51		40	1.57
14	0.55		41	1.61
15	0.59		42	1.65
16	0.63		43	1.69
17	0.67		44	1.73
18	0.71		45	1.77
19	0.75		46	1.81
20	0.79		47	1.85
21	0.83		48	1.89
22	0.87		49	1.93
23	0.91		50	1.97
24	0.94		51	2.01
25	0.98		52	2.05
26	1.02		53	2.09
27	1.06		54	2.13

Fractional Inches	Decimal Inches	mm
1/32	0.03	0.8
1/16	0.06	1.6
3/32	0.09	2.4
1/8	0.13	3.2
5/32	0.16	4.0
3/16	0.19	4.8
7/32	0.22	5.6
1/4	0.25	6.4
9/32	0.28	7.1
5/16	0.31	7.9
11/32	0.34	8.7
3/8	0.38	9.5
13/32	0.41	10.3
7/16	0.44	11.1
15/32	0.47	11.9
1/2	0.50	12.7
17/32	0.53	13.5
9/16	0.56	14.3
19/32	0.59	15.1
5/8	0.63	15.9
21/32	0.66	16.7
11/16	0.69	17.5
23/32	0.72	18.3
3/4	0.75	19.1
25/32	0.78	19.8
13/16	0.81	20.6
27/32	0.84	21.4
7/8	0.88	22.2
29/32	0.91	23.0
15/16	0.94	23.8
31/32	0.97	24.6
1	1.00	25.4

Appendix H
Hook Terminology

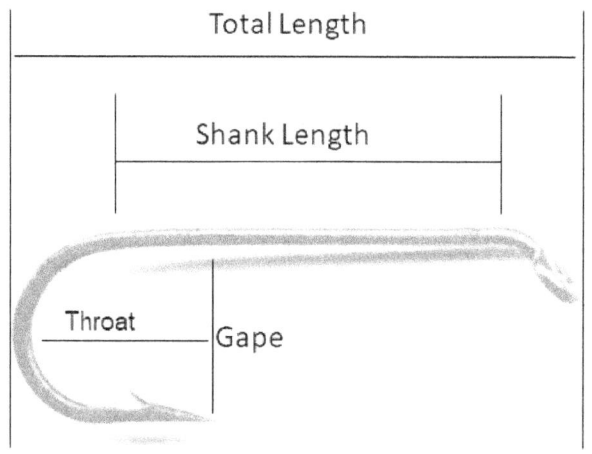

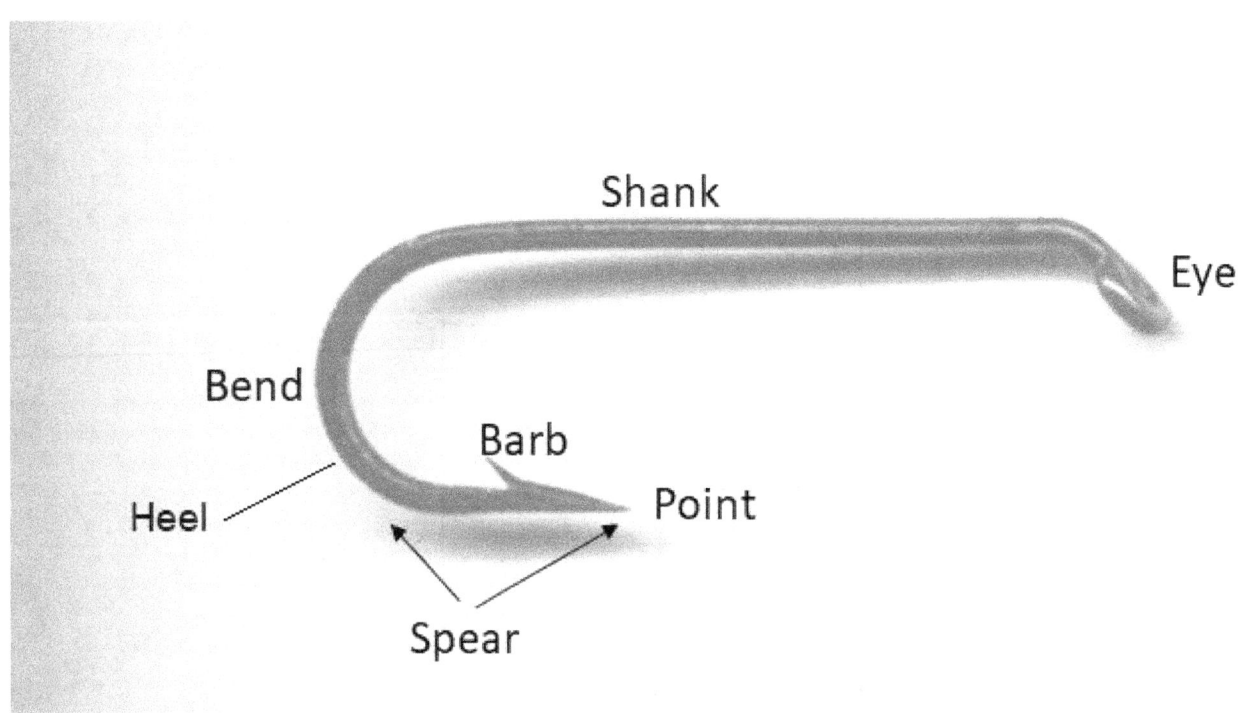

Appendix I
Fly Proportions and Nomenclature

Proportions of a Traditional Catskill Style Dry Fly

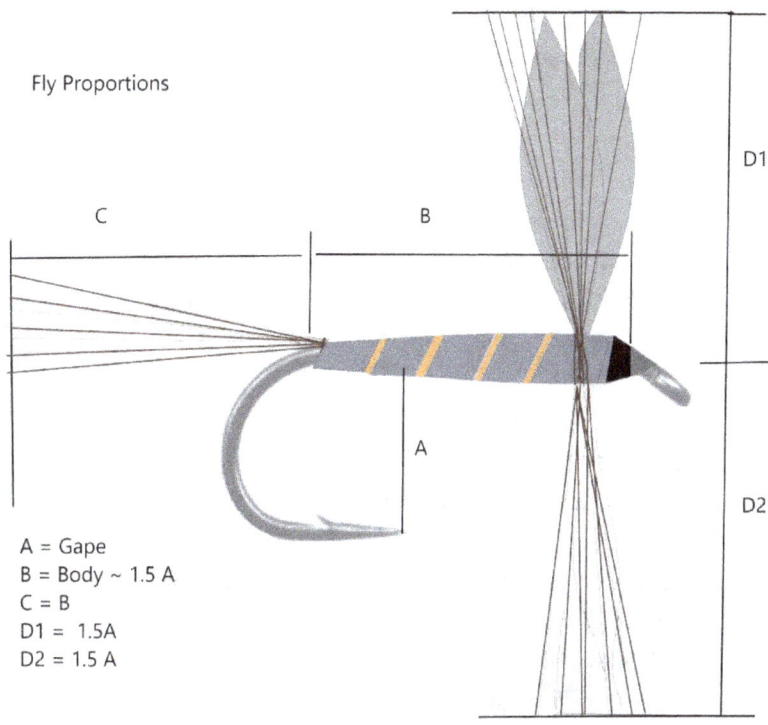

Fly Proportions

C

B

D1

A

D2

A = Gape
B = Body ~ 1.5 A
C = B
D1 = 1.5A
D2 = 1.5 A

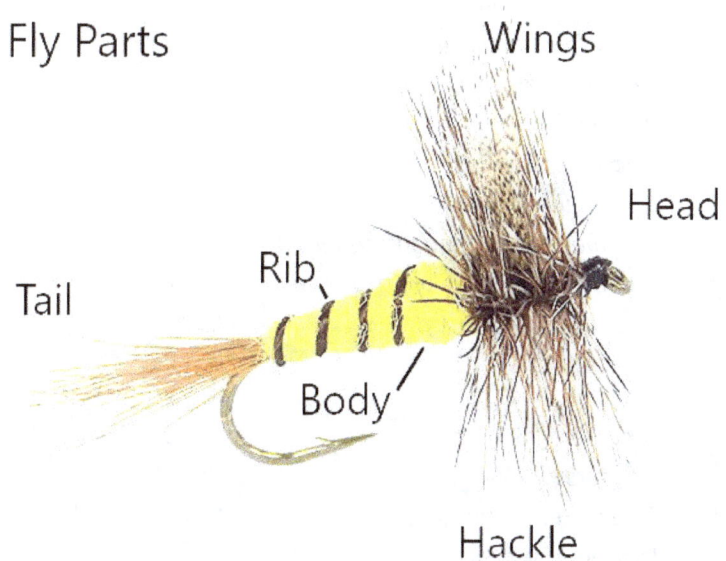

Fly Parts

Wings

Head

Rib

Tail

Body

Hackle

Parts of a Traditional Catskill Style Dry Fly

Appendix J
Insects

Mayflies

MAYFLIES Common Name	Genus & species	Adult Body mm
Blue Quill	*Leptophlebia cupida*	10 - 12
Blue Quill	*Paraleptophlebia adoptiva*	6 - 8
Blue Winged Olive	*Attenella attenuata*	6 - 8
Blue Winged Olive	*Baetis tricaudatus*	6 - 8
Blue Winged Olive	*Callibaetus ferrugineus*	7 - 9
Blue Winged Olive	*Drunella cornuta*	9 - 10
Blue Winged Olive	*Drunella lata*	6 - 8
Blue Winged Olive	*Iswaeon anoka*	4 - 5
Brown Drake	*Ephemera simulans*	11 - 14
Ginger Quill	*Stenacron interpunctatum*	9 - 11
Golden Drake	*Anthopotamus distinctus*	13 - 16
Gray Drake	*Siphlonurus alternatus*	12 - 16
Gray Fox	*Maccaffertium vicarium*	12 - 14
Green Drake	*Ephemera guttalata*	18 - 30
Hendrickson	*Ephemerella subvaria*	10 - 12
Hex	*Hexagenia atrocaudata*	16 - 35
Hex	*Hexagenia rigida*	16 - 35
Light Cahill	*Maccaffertium ithaca*	10 - 12
Mahogany Dun	*Isonychia bicolor*	13 - 16
March Brown	*Maccaffertium vicarium*	14 - 16
Michigan Caddis (Hex)	*Hexagenia limbata*	16 - 35
Pale Evening Dun	*Epeorus (sp)*	9 - 12
Pale Evening Dun	*Maccaffertium (sp)*	6 - 12
Pale Evening Dun	*Rithrogena (sp)*	7 - 11
Pale Morning Dun	*Ephemerella dorthea infrequens*	7 - 8
Pale Morning Dun	*Ephemerella excrucians*	5 - 7
Quill Gordon	*Epeorus pleuralis*	9 - 12
Slate Drake / LW Coachman	*Isonychia bicolor*	13 - 16
Sulphur	*Ephemerella dorthea*	6 - 8
Sulphur	*Ephemerella invaria*	8 - 9
Sulphur	*Ephemerella invaria*	9 - 11
Trico	*Tricorythodes (sp)*	3 - 4
Western Green Drake	*Drunella doddsi*	12 - 13
Western Green Drake	*Drunella grandis*	13 - 15
Western Green Drake	*Drunella grandis*	14 - 16
Western Slate Wing Olive	*Drunella flavilinea*	8 - 10

| White Fly | *Ephoron leukon* | 13 - 14 |
| Yellow Drake | *Ephemera varia* | 13 - 16 |

Stoneflies

Common Name	Family	Adult Body mm
Golden Stoneflies	*Perlidae*	9 - 40
Green Stoneflies	*Chloroperlidae*	6 - 15
Stripetails	*Perlodidae*	10 - 15
Roach-like Stoneflies	*Peltoperlidae*	5 - 10
Salmonflies or Giant Stoneflies	*Pteronarcyidae*	25 - 60

Caddisflies

Common Name	Family	Adult Body mm
Fingernet Caddisflies	*Philopotamidae*	5 -9
Tube Maker Caddisflies	*Polycentropodidae*	8 -25
Net Tube Caddisflies	*Psychomyiidae*	8 -25
Net Spinning Caddisflies	*Hydropsychidae*	6 -19
Little Black Caddisflies	*Glossosomatidae*	5 -10
Microcaddisflies	*Hydroptilidae*	3 -10
Free-Living Caddisflies	*Rhyacophilidae*	10 -50
Hood Casemakers	*Molannidae*	8 -25
Long Horn Caddisflies	*Leptoceridae*	5 -17
Mortarjoint Caddisflies	*Odontoceridae*	7 -25
Early Smoky Wing Sedges	*Apataniidae*	7 -25
Northern Caddisflies	*Limnephilidae*	7 -23
Square Wood Case Caddisflies	*Brachycentridae*	7 -18
Log Cabin Caddisflies	*Lepidostomatidae*	5 -15
Giant Casemakers	*Phryganeidae*	18 -30

Appendix K
How Hook Sizes Were Arrived At

As mentioned previously hooks makers are a secretive lot. They are also competitive. Sizes are generally agreed on now as low number sizes are larger than high numbers. That is to say a size 2 is a larger hook than a size 12 which is larger than a size 20. Though through time there have been scales devised where higher numbers are larger. But there is no standard for all hook sizes from all companies. There does not need to be. One possible explanation for hook sizes follows.

Picture if you will a blacksmith shop in old England with the sign above the door that read:

Howard, Fine and Howard
Blacksmiths
Nobody can beat our steel

Curly and Larry stand over a glowing forge, roasting wieners on metal rods. Larry takes a bite out of his and makes a face. "Say this hotdog doesn't taste so hot."

Curly takes a bite out of his and chomps off a piece of the metal rod. As he chews there is a crunching sound. "Hmmm. Mine's okay but a little crunchy." He continues to chew and takes another bite. As he goes to take the last bite, he realizes about 4 inches is missing from the rod. He pulls his hand down over his face. "Nyaaahh!"

In walks Moe. "Listen up you knuckleheads. The king wants to go a fishing and he needs hooks. Larry, get busy on the bellows. Start pounding some steel Curly."

Moe turns to watch Larry pump the bellows and with each pull down on the long wooden lever a blast of flame emanates from the forge singeing his backside.

Curly picks up a piece of iron from the forge and begins to pound it on the anvil. "La da dee, la dee, la dee."

Moe turns to him, "Enough with the singing. Get busy!" He takes the hammer from Curly and hits him in the head with a resounding gong.

"Whoa Moe. That was uncalled for," says Curly.

"I'll show you uncalled for." He picks up a pair of tongs and twists Curly's nose. When he lets go, he looks at the tongs and they are bent. "Hmmph, need new tongs. Okay you two meatheads get to work. The king needs the hooks. I'll be back in a little while."

Moe leaves and Curly and Larry continue to work, with Larry pumping the bellows and flame shooting out of the forge. Curly looks at the forge. "Rrrrrrrufffff!"

"Roaaaaar!" bellows the forge as flams come out and singe Curly's face. "Nyaaaahhh!"

A while later Curly holds up a fish hook. "It poyfect! A work of art!"

Larry is standing beside him belching. "That wiener didn't agree with me," he says pounding his sternum.

In walks Moe. "Are the kings hooks ready?"

"Sointently," says Curly holding up his creation.

"You just made one?" asks Moe. "I said the king needs hooks. That's plural. That means more than one dummy. Now get busy or we'll all be in the royal dungeon."

Curly picks up another piece of iron and begins to pound, while Larry alternates pumping the bellows and belching. With each belch the flame in the forge roars.

A while later Curly produces a second hook, this one smaller than the rest. "Another poifection."

"Is that one too? Asks Moe.

"It's two," says Curly "Nyuk, nyuk, nuyk."

"What size is it?" asks Moe.

"It's a hook too," says Curly beaming.

"It's a two?"

"It is too."

"So we have a one and a two?"

Larry intervenes, "A one and a two."

Curly takes his hand and they begin to dance. "La la la la la la la la la la"

Moe bounces their heads together. Boink! "Cut it out you knuckleheads. The king will be here any minute."

In walks the king, a large stately man wearing a crown and a fishing vest. "Are my fishing hooks ready? I have an Itchen to get on with the Test."

"They're right here kingsy wingsy," says Curly holding up the hooks.

The king takes the hooks and begins to inspect them. "What sizes are they?"

"Well your anglership, that's one," says Moe.

"And the other?"

"It's one too," says Curly rocking back and forth, "Nyuk, nyuk, nuyk."

"Very well," says the king, "I decree that this," holding up the larger hook, "This is size one. And this," holding up the smaller one, "is size two."

The king looks closely at the smaller hook's bend. "That bend is unique. What do you call it?"

Larry pounding his stomach lets go a belch, "Sproaaaat!" The flames in the forge flare wildly.

"Very well. I decree it will be a Sproat bend," says the king. He then picks up the smaller hook and inspects it, "This one has a different bend. What do you call it?

Larry runs over to a barrel and sticks his head in it, "Yoooooorrrrrrrk!" Water splashes out of the barrel.

"I decree that this bend shall henceforth be known as a York bend."

Moe pulls Larry out of the barrel by his hair and yanks out a hank in the process. Looking at it you can see he is deep in thought, then he hands it to Curly. "Here knucklehead, make some hooks out of this."

"I can and they will be mighty Fine hooks. Nyuk, nyuk, nuyk." Then pulling another hank of hair from Larry, "And these will be an extra Fine hook." And thus, the designation of Fine and Extra Fine came about.

The Three Blind Mice theme song comes on and fade to black.

Well maybe it didn't happen that way but do you have a better story?

Life List
Based on International Game Fish Association Fish Records
Note that Multiple Categories of Records May Exist for Each Species

Common Name	Scientific Name	Cgt	Fly
Acara Paragaio	*Hoplarchus psittacus*		
Acara Roi Roi	*Geophagus proximus*		
Acara Tucanare	*Satanoperca lilith*		
Acara, Silver	*Chaetobranchus flavescens*		
Aimara	*Hoplerythrinus unitaeniatus*		
Akahata	*Epinephelus fasciatus*		
Aka-Matsu	*Deoderleinia berycoides*		
Akodai	*Sebastes matsubarai*		
Albacore	*Thunnus alalunga*		
Alfonsino	*Beryx decadactylus*		
Alfonsino, Splendid	*Beryx splendens*		
Altai Osman	*Oreoleuciscus potanini*		
Amberjack, Greater	*Seriola dumerili*		
Angelfish, Grey	*Pomacanthus arcuatus*		
Angelfish, King	*Holacanthus passer*		
Angelfish, Semicircle	*Pomacanthus semicirculatus*		
Angelshark, Japanese	*Squatina japonica*		
Angler	*Lophius piscatorius*		
Ara	*Niphon spinosus*		
Araçu Comum	*Schizodon vittatus*		
Aracu	*Leporinus agassizii*		
Aracu, Fat Head	*Leporinus trifasciatus*		
Aracu, Pau-de-nego	*Rhytiodus microlepis*		
Arapaima	*Arapaima gigas*		
Arawana	*Osteoglossum bicirrhosum*		
Arawana, Black	*Osteoglossum ferreirai*		

Common Name	Scientific Name		
Argentine, Greater	*Argentina silus*		
Asp	*Aspius aspius*		
Ayamekasago	*Sebastiscus albofasciatus*		
Barb, Boney Lipped	*Osteochilus melanopleurus*		
Barb, Eye-spot	*Hampala dispar*		
Barb, Giant	*Catlocarpio siamensis*		
Barb, Golden Belly	*Hypsibarbus wetmorei*		
Barb, Golden Tinfoil	*Hypsibarbus malcolmi*		
Barb, Hampala	*Hampala macrolepidota*		
Barb, Java	*Barbonymus gonionotus*		
Barb, Lagleri	*Hypsibarbus lagleri*		
Barb, Laos	*Poropuntius laoensis*		
Barb, Mad	*Leptobarbus hoevenii*		
Barb, Smith's	*Puntioplites proctozystron*		
Barb, Soldier River	*Cyclocheilichthys enoplos*		
Barb, Tinfoil	*Barbonymus schwanenfeldii*		
Barb, Two-barred	*Hampala bimaculata*		
Barbel	*Barbus barbus*		
Barbel, Andalusian	*Luciobarbus sclateri*		
Barbel, Comizo Hybrid	*Luciobarbus steindachner*		
Barbel, Iberian	*Luciobarbus bocagei*		
Barbel, Steed	*Hemibarbus labeo*		
Barracuda, Bigeye	*Sphyraena forsteri*		
Barracuda, Blackfin	*Sphyraena qenie*		
Barracuda, Great	*Sphyraena barracuda*		
Barracuda, Guinean	*Sphyraena afra*		
Barracuda, Hellers	*Sphyraena helleri*		

Barracuda, Mexican	Sphyraena ensis		
Barracuda, Pacific	Sphyraena argentea		
Barracuda, Pelican	Sphyraena idiastes		
Barracuda, Pickhandle	Sphyraena jello		
Barracuda, Yellowmouth	Sphyraena viridensis		
Barramundi	Lates calcarifer		
Barrelfish	Hyperoglyphe perciformis		
Barrelfish, Pacific	Hyperoglyphe japonica		
Baru	Uaru amphiacanthoides		
Bass, Alabama	Micropterus henshalli		
Bass, Australian	Macquaria novemaculeata		
Bass, Barred Sand	Paralabrax nebulifer		
Bass, Black Sea	Centropristis striata		
Bass, Damsel	Hemanthias signifer		
Bass, European	Dicentrarchus labrax		
Bass, Giant Sea	Stereolepis gigas		
Bass, Goldspotted Sand	Paralabrax auroguttatus		
Bass, Guadalupe	Micropterus treculii		
Bass, Guadalupe X Smallmouth	Micropterus treculi x M. dolomieu		
Bass, Kelp	Paralabrax clathratus		
Bass, Largemouth	Micropterus salmoides		
Bass, Leather	Dermatolepis dermatolepis		
Bass, Longtail	Hemanthias leptus		
Bass, Meanmouth	Micropterus dolomieu x M. punctulatus		
Bass, Ozark	Ambloplites constellatus		
Bass, Roanoke	Ambloplites cavifrons		
Bass, Rock	Ambloplites rupestris		
Bass, Shadow	Ambloplites ariommus		
Bass, Smallmouth	Micropterus dolomieu		

Bass, Smallmouth X Alabama	Micropterus dolomieu x henshalli		
Bass, Smallmouth X Largemouth	Micropterus dolomieu x salmoides		
Bass, Spotted	Micropterus punctulatus		
Bass, Spotted Sand	Paralabrax maculatofasciatus		
Bass, Striped	Morone saxatilis		
Bass, Striped (landlocked)	Morone saxatilis		
Bass, Suwannee	Micropterus notius		
Bass, White	Morone chrysops		
Bass, Whiterock	Morone saxatilis x M. chrysops		
Bass, Yellow	Morone mississippiensis		
Bass, Yellow (hybrid)	Morone mississippiensis x M. chrysops		
Batfish, Orbicular	Platax orbicularis		
Batfish, Tiera	Platax teira		
Beardfish, Pacific	Polymixia berndti		
Biara	Rhaphiodon vulpinus		
Bicuda	Boulengerella cuvieri		
Bigeye	Priacanthus arenatus		
Bigeye, Short	Pristigenys alta		
Binga	Dimidiochromis kiwinge		
Blackfish, Sacramento	Orthodon microlepidotus		
Blackfish, Smallscale	Girella punctata		
Blennie, Patagonian	Eleginops maclovinus		
Bludger	Carangoides gymnostethus		
Bluefish	Pomatomus saltatrix		
Bluegill	Lepomis macrochirus		
Bluegill/Sunfish, Redear (hybrid)	Lepomis macrochirus x L. microlophus		
Bocaccio	Sebastes paucispinis		

2

Boga	*Leporinus obstusidens*				Bream, White	*Blicca bjoerkna*			
Bonefish	*Albula vulpes*				Bream, Yellowfin Sea	*Dentex abei*			
Bonefish, Cortez	*Albula gilberti*				Brill	*Scophthalmus rhombus*			
Bonefish, Roundjaw	*Albula glossodonta*				Brotula, Bearded	*Brotula barbata*			
Bonefish, Sharpjaw	*Albula virgata*				Brotula, Pacific Bearded	*Brotula clarkae*			
Bonefish, Smallscale	*Albula oligolepis*				Brycon	*Brycon amazonicus*			
Boneytongue, Australian	*Scleropages jardinii*				Buffalo, Bigmouth	*Ictiobus cyprinellus*			
Bonito, Atlantic	*Sarda sarda*				Buffalo, Black	*Ictiobus niger*			
Bonito, Australian	*Sarda australis*				Buffalo, Smallmouth	*Ictiobus bubalus*			
Bonito, Leaping	*Cybiosarda elegans*				Bulleye, Arrow	*Priacanthus sagittarius*			
Bonito, Pacific	*Sarda lineolata*				Bullhead, Black	*Ameiurus melas*			
Bonito, Striped	*Sarda orientalis*				Bullhead, Brown	*Ameiurus nebulosus*			
Bowfin	*Amia calva*				Bullhead, Yellow	*Ameiurus natalis*			
Boxfish, Bluespotted	*Ostracion immaculatus*				Bullseye, Longfinned	*Cookeolus japonicus*			
Boxfish, Yellow	*Ostracion cubicum*				Bullseye, Moontail	*Priacanthus hamrur*			
Bracanjuva	*Brycon orbignyanus*				Burbot	*Lota lota*			
Bream	*Abramis brama*				Buri (Japanese Amberjack)	*Seriola quinqueradiata*			
Bream, African Red	*Pagrus africanus*				Burrfish, Spotfin	*Chilomycterus reticulatus*			
Bream, Black	*Hephaestus fuliginosus*				Burrfish, Spotted	*Chilomycterus reticulatus*			
Bream, Blue-lined Large-eye	*Gymnocranius grandoculis*				Burrfish, Striped	*Chilomycterus schoepfii*			
Bream, Blue-spotted Large-eye	*Gymnocranius microdon*				Butterfish, Brazilian	*Hyperoglyphe macrophthalma*			
Bream, Collared Large-eye	*Gymnocranius audleyi*				Butterflyfish, Lined	*Chaetodon lineolatus*			
Bream, Eyebrowed Large Eye	*Gymnocranius superciliosus*				Cabezon	*Scorpaenichthys marmoratus*			
Bream, Forktail Large-eye	*Gymnocranius elongatus*				Cachorro, Peixe	*Acestrorhynchus falcirostris*			
Bream, Grey Large-eye	*Gymnocranius griseus*				Carp, Bighead	*Hypophthalmichthys nobilis*			
Bream, Humpnose Big-eye	*Monotaxis grandoculis*				Carp, Black	*Mylopharyngodon piceus*			
Bream, Japanese Large-eye	*Gymnocranius euanus*				Carp, Common	*Cyprinus carpio*			
Bream, King Soldier	*Argyrops spinifer*				Carp, Crucian	*Carassius carassius*			
Bream, Mozambique Large-eye	*Wattsia mossambica*				Carp, Grass	*Ctenopharyngodon idella*			
Bream, Rosy Dwarf Monocle	*Parascolopsis eriomma*				Carp, Predatory	*Chanodichthys erythropterus*			
Bream, Threadfin Yellowbelly	*Nemipterus bathybius*				Carp, Silver	*Hypophthalmichthys molitrix*			
Bream, Twoband	*Diplodus vulgaris*								

Common Name	Scientific Name			Common Name	Scientific Name		
Carpsucker, River	*Carpiodes carpio*			Catfish, Smoothmouth Sea	*Carlarius heudelotii*		
Catalufa, Popeye	*Pristigenys serrula*			Catfish, Swai	*Pangasianodon hypophthalmus*		
Catfish, Amur	*Silurus asotus*			Catfish, Thai Shark	*Helicophagus leptorhynchus*		
Catfish, Asian Redtail	*Hemibagrus nemurus*			Catfish, Thickspined	*Nemapteryx nenga*		
Catfish, Black	*Hemibagrus wyckii*			Catfish, Tigerstriped	*Brachyplatystoma tigrinum*		
Catfish, Blue	*Ictalurus furcatus*			Catfish, Vermiculated Sailfin	*Pterygoplichthys disjunctivus*		
Catfish, Bronze	*Netuma bilineata*			Catfish, Walking	*Clarias batrachus*		
Catfish, Channel	*Ictalurus punctatus*			Catfish, White	*Ameiurus catus*		
Catfish, Duckbill	*Sorubim lima*			Catfish, White Sea	*Genidens barbus*		
Catfish, Eurasian	*Silurus biwaensis*			Catfish, Wyckioides	*Hemibagrus wyckioides*		
Catfish, Firewood	*Sorubimichthys planiceps*			Catfish, Yellow	*Pimelodus maculatus*		
Catfish, Flapnose Sea	*Sciades dowii*			Catfish, Zebra	*Brachyplatystoma juruense*		
Catfish, Flathead	*Pylodictis olivaris*			Catla	*Catla catla*		
Catfish, Flatwhiskered	*Pinirampus pirinampu*			Catshark, Brownbanded Bamboo	*Chiloscyllium punctatum*		
Catfish, Gafftopsail	*Bagre marinus*			Catshark, Small-spotted	*Scyliorhinus canicula*		
Catfish, Giant	*Netuma thalassina*			Chalceus, Pinktail	*Chalceus macrolepidotus*		
Catfish, Giant (Mekong)	*Pangasianodon gigas*			Char, Arctic	*Salvelinus alpinus*		
Catfish, Gilded	*Brachyplatystoma rousseauxii*			Char, Sparctic	*Salvelinus alpinus x fontinalis*		
Catfish, Granulated	*Pterodoras granulosus*			Char, Whitespotted	*Salvelinus leucomaenis*		
Catfish, Grey Eel	*Plotosus canius*			Chilipepper	*Sebastes goodei*		
Catfish, Hardhead	*Ariopsis felis*			Chinamanfish	*Symphorus nematophorus*		
Catfish, Highwaterman	*Hypothalmus edentatus*			Chiselmouth	*Acrocheilus alutaceus*		
Catfish, Mandi	*Pimelodus ornatus*			Chub	*Squalius cephalus*		
Catfish, Orinoco Sailfin	*Pterygoplichthys multiradiatus*			Chub, Bermuda	*Kyphosus sectatrix*		
Catfish, Pangasius	*Pangasius mekongensis*			Chub, Blue Sea	*Kyphosus cinerascens*		
Catfish, Redtail	*Phractocephalus hemioliopterus*			Chub, Brassy	*Kyphosus vaigiensis*		
Catfish, Ripsaw	*Oxydoras niger*			Chub, Cortez Sea	*Kyphosus elegans*		
Catfish, Sagor	*Hexanematichthys sagor*			Chub, Grey Sea	*Kyphosus bigibbus*		
Catfish, Sharptooth	*Clarias gariepinus*			Chub, Roundtail	*Gila robusta*		
Catfish, Slobbering	*Brachyplatystoma platynemum*			Chub, Tui	*Siphateles bicolor*		

Chub, Utah	*Gila Atraria*			Conger, Mulatto	*Enchelycore nigricans*		
Chub, Yellow	*Kyphosus incisor*			Conger, Philippine	*Conger philippinus*		
Chubsucker, Eastern Creek	*Erimyzon oblongus*			Conger, Red Pike	*Cynoponticus coniceps*		
Cichild, Redhead	*Vieja melanurus*			Conger, Silvery	*Ariosoma meeki*		
Cichlid, Banded	*Heros severus*			Coralgrouper, Blacksaddled	*Plectropomus laevis*		
Cichlid, Blackbelt	*Vieja maculicauda*			Coralgrouper, Highfin	*Plectropomus oligacanthus*		
Cichlid, Giant Tanganyika	*Boulengerochromis microlepis*			Coralgrouper, Leopard	*Plectropomus leopardus*		
Cichlid, Mayan	*Mayaheros urophthalmus*			Coralgrouper, Roving	*Plectropomus pessuliferus*		
Cichlid, Melanura	*Paraneetroplus melanura*			Coralgrouper, Squaretail	*Plectropomus areolatus*		
Cichlid, Midas	*Amphilophus citrinellus*			Corb	*Umbrina cirrosa*		
Cichlid, Oaxaca	*Vieja zonata*			Corbina, California	*Menticirrhus undulatus*		
Cichlid, Redhead	*Vieja melanura*			Coris, Clown	*Coris aygula*		
Cichlid, Rio Grande	*Herichthys cyanoguttatus*			Coris, Yellowstripe	*Coris flavovittata*		
Cichlid, Turquoise	*Caquetaia umbrifera*			Cornetfish	*Fistularia tabacaria*		
Cisco	*Coregonus artedi*			Cornetfish, Bluespotted	*Fistularia commersonii*		
Cobia	*Rachycentron canadum*			Cornetfish, Red	*Fistularia petimba*		
Cod, Atlantic	*Gadus morhua*			Coroata	*Platynematichthys notatus*		
Cod, Cow	*Sebastes levis*			Corvina, Hybrid	*Cynoscion xanthulus x C. nebulosus*		
Cod, New Zealand Blue	*Parapercis colias*			Corvina, Orangemouth	*Cynoscion xanthulus*		
Cod, Pacific	*Gadus macrocephalus*			Corvina, Shortfin	*Cynoscion parvipinnis*		
Cod, Red Rock	*Scorpaena cardinalis*			Corvina, Whitefin	*Cynoscion albus*		
Cod, Saffron	*Eleginus gracilis*			Cowfish, Roundbelly	*Lactoria diaphana*		
Codling, Brazilian	*Urophycis brasiliensis*			Crappie, Black	*Pomoxis nigromaculatus*		
Codling, Japanese	*Physiculus japonicus*			Crappie, White	*Pomoxis annularis*		
Codling, Southern Bastard	*Pseudophycis barbata*			Creek Chub	*Semotilus atromaculatus*		
Codling, Spotted	*Urophycis regia*			Creole-fish	*Paranthias furcifer*		
Coney	*Cephalopholis fulva*			Creolefish, Pacific	*Paranthias colonus*		
Coney, Gulf	*Hyporthodus acanthistius*			Croaker, Atlantic	*Micropogonias undulatus*		
Conger	*Conger conger*			Croaker, Black	*Cheilotrema saturnum*		
Conger, Argentine	*Conger orbignianus*			Croaker, Black	*Chielotrema saturnum*		
Conger, Cape	*Conger wilsoni*						
Conger, Hawaiian Mustache	*Conger marginatus*						
Conger, Japanese	*Conger japonicus*						

Croaker, Blackmouth	*Atrobucca Nibe*				Dogfish, Picked	*Squalus acanthias*		
Croaker, Boeseman	*Boesemania microlepis*				Dogfish, Rough Longnose	*Deania hystricosa*		
Croaker, Camaroon	*Pseudotolithus moorii*				Dogfish, Roughskin	*Centroscymnus owstonii*		
Croaker, Cassava	*Pseudotolithus senegalensis*				Dogfish, Smooth	*Mustelus canis*		
Croaker, Honnibe	*Nibea mitsukurii*				Dogfish, Whitetail	*Squalus albicaudus*		
Croaker, Law	*Pseudotolithus senegallus*				Dolly Varden	*Salvelinus malma*		
Croaker, Longneck	*Pseudotolithus typus*				Dolphinfish	*Coryphaena hippurus*		
Croaker, S. A. Silver	*Plagioscion squamosissimus*				Dolphinfish, Pompano	*Coryphaena equiselis*		
Croaker, Silver	*Pennahia argentata*				Dorada	*Brycon moorei*		
Croaker, Spotfin	*Roncador stearnsii*				Dorado	*Salminus brasiliensis*		
Croaker, White	*Genyonemus lineatus*				Dory, Mirror	*Zenopsis nebulosa*		
Croaker, Whitemouth	*Micropogonias furnieri*				Dourado	*Salminus hilarii*		
Croaker, Yellowfin	*Umbrina roncador*				Drum, Black	*Pogonias cromis*		
Cui-ui	*Chasmistes cujus*				Drum, Canary	*Umbrina canariensis*		
Cunner	*Tautogolabrus adspersus*				Drum, Freshwater	*Aplodinotus grunniens*		
Curbinata, Black	*Plagioscion auratus*				Drum, Polla	*Umbrina xanti*		
Dab	*Limanda limanda*				Drum, Red	*Sciaenops ocellatus*		
Dab, Long Rough	*Hippoglossoides platessoides*				Drum, Yellow	*Nibea albiflora*		
Dainan-anago	*Conger erebennus*				Drummer, Black	*Girella elevata*		
Dainanumihebi	*Ophisurus macrorhynchos*				Drummer, Pacific	*Kyphosus pacificus*		
Dart, Largespotted	*Trachinotus botla*				Durgon, Black	*Melichthys niger*		
Dart, Small Spotted	*Trachinotus baillonii*				Eel, American	*Anguilla rostrata*		
Dentex	*Dentex dentex*				Eel, Conger	*Conger oceanicus*		
Dentex, Canary	*Dentex canariensis*				Eel, Cutthroat Grey	*Synaphobranchus affinis*		
Dentex, Pink	*Dentex gibbosus*				Eel, European	*Anguilla anguilla*		
Doctorfish	*Acanthurus chirurgus*				Eel, Fire	*Mastacembelus erythrotaenia*		
Dogfish, Birdbeak	*Deania calcea*				Eel, Highfin-snake	*Opichthus zophistius*		
Dogfish, Hawaiian	*Squalus hawaiiensis*				Eel, Japanese	*Anguilla japonica*		
Dogfish, Mandarin	*Cirrhigaleus barbifer*				Eel, Kaup's Arrowtooth	*Synaphobranchus kaupii*		
Dogfish, Northern Spiny	*Squalus griffini*				Eel, King Snake	*Ophichthus rex*		
Dogfish, Pacific Spiny	*Squalus suckleyi*				Eel, Leopard Moray	*Enchelycore pardalis*		

Eel, Marbled	*Anguilla marmorata*		
Eel, Shortfin	*Anguilla australis*		
Eel, Spotted Moray	*Gymnothorax isingteena*		
Eel, Stippled Spoon-nose	*Echiophis punctifer*		
Eel, Tiger Reef	*Scuticaria tigrina*		
Emperor, Drub	*Lethrinus ravus*		
Emperor, Longface	*Lethrinus olivaceus*		
Emperor, Pacific Yellowtail	*Lethrinus atkinsoni*		
Emperor, Pink Ear	*Lethrinus lentjan*		
Emperor, Sky	*Lethrinus mahsena*		
Emperor, Spangled	*Lethrinus nebulosus*		
Emperor, Spotcheek	*Lethrinus rubrioperculatus*		
Emperor, Thumbprint	*Lethrinus harak*		
Emperor, Yellowlip	*Lethrinus xanthochilus*		
Escolar	*Lepidocybium flavobrunneum*		
Escolar, Roudi	*Promethichthys prometheus*		
Escolar, Royal	*Rexea prometheoides*		
Ezo-Mebaru	*Sebastes taczanowskii*		
Fallfish	*Semotilus corporalis*		
Fanray	*Platyrhina sinensis*		
Featherback, Clown	*Chitala ornata*		
Featherback, Giant	*Chitala lopis*		
Featherback, Indochina	*Chitala blanci*		
Filefish, Modest	*Thamnaconus modestoides*		
Filefish, Scrawled	*Aluterus scriptus*		
Filefish, Thread-sail	*Stephanolepis cirrhifer*		
Filefish, Unicorn	*Aluterus monoceros*		
Filefish, Whitespotted	*Cantherhines dumerilii*		
Firefish, Devil	*Pterois miles*		
Flagfin, Royal	*Aulopus filamentosus*		

Flagtail, Rock	*Kuhlia rupestris*		
Flathead, Bar-tailed	*Platycephalus indicus*		
Flathead, Crocodile	*Cociella crocodilus*		
Flathead, Dusky	*Platycephalus fuscus*		
Flathead, Spot Eye	*Inegocia ochiaii*		
Flier	*Centrarchus macropterus*		
Flounder, Arrow Tooth	*Atheresthes stomias*		
Flounder, Brazilian	*Paralichthys brasiliensis*		
Flounder, Cinnamon	*Pseudorhombus cinnamoneus*		
Flounder, Cresthead	*Pseudopleuronectes schrenki*		
Flounder, European	*Platichthys flesus*		
Flounder, Far Eastern Smooth	*Liopsetta pinnifasciata*		
Flounder, Gulf	*Paralichthys albigutta*		
Flounder, Marbled	*Pseudopleuronectes yokohamae*		
Flounder, Ocellated	*Pseudorhombus dupliciocellatus*		
Flounder, Olive	*Paralichthys olivaceus*		
Flounder, Sand	*Limanda punctatissima*		
Flounder, Slime	*Microstomus achne*		
Flounder, Small-eyed	*Paralichthys microps*		
Flounder, Southern	*Paralichthys lethostigma*		
Flounder, Speckled	*Paralichthys woolmani*		
Flounder, Starry	*Platichthys stellatus*		
Flounder, Stone	*Kareius bicoloratus*		
Flounder, Summer	*Paralichthys dentatus*		
Flounder, Toothed	*Cyclopsetta querna*		
Flounder, Winter	*Pseudopleuronectes americanus*		
Flounder, Yellow Striped	*Pseudopleuronectes herzensteini*		
Flounder, Yellowbelly	*Rhombosolea leporina*		
Forkbeard	*Phycis phycis*		
Forkbeard, Greater	*Phycis blennoides*		

Frogfish, Ocellated	*Fowlerichthys ocellatus*			Goatfish, Yellowstripe	*Mulloidichthys flavolineatus*		
Fusilier, Blue And Gold	*Caesio caerulaurea*			Goby, Marble	*Oxyeleotris marmorata*		
Fusilier, Redbelly Yellowtail	*Caesio cuning*			Goldeye	*Hiodon alosoides*		
Gar, Alligator	*Atractosteus spatula*			Goldfish	*Carassius auratus*		
Gar, Florida	*Lepisosteus platyrhincus*			Goldfish, Asian	*Carassius auratus langsdorfii*		
Gar, Hybrid	*Lepisosteus osseus x Atractosteus spatula*			Goldfish-Carp Hybrid	*Carassius auratus x Cyprinus carpio*		
Gar, Longnose	*Lepisosteus osseus*			Goonch	*Bagarius bagarius*		
Gar, Shortnose	*Lepisosteus platostomus*			Goosefish	*Lophius americanus*		
Gar, Spotted	*Lepisosteus oculatus*			Goosefish, Yellow	*Lophius litulon*		
Gar, Tropical	*Atractosteus tropicus*			Gourami, Elephant Ear	*Osphronemus exodon*		
Garpike	*Belone belone*			Gourami, Giant	*Osphronemus goramy*		
Geelbek	*Atractoscion aequidens*			Gourami, Giant Borneo	*Osphronemus septemfasciatus*		
Gemfish, Silver	*Rexea solandri*			Grayling	*Thymallus thymallus*		
Gengoro-Buna	*Carassius cuvieri*			Grayling, Arctic	*Thymallus arcticus*		
Gissu, Japanese	*Pterothrissus gissu*			Grayling, Mongolian	*Thymallus brevirostris*		
Globe Fish	*Tetraodon lineatus*			Graysby	*Cephalopholis cruentata*		
Gnomefish (mutsu)	*Scombrops boops*			Graysby, Panama	*Cephalopholis panamensis*		
Goatfish, Blackspot	*Parupeneus spilurus*			Greenling, Fat	*Hexagrammos otakii*		
Goatfish, Cinnabar	*Parupeneus heptacanthus*			Greenling, Kelp	*Hexagrammos decagrammus*		
Goatfish, Doublebar	*Parupeneus trifasciatus*			Greenling, Masked	*Hexagrammos octogrammus*		
Goatfish, Gold-saddle	*Parupeneus cyclostomus*			Greenling, Rock	*Hexagrammos lagocephalus*		
Goatfish, Indian	*Parupeneus indicus*			Grenadier, Japanese	*Coelorinchus japonicus*		
Goatfish, Manybar	*Parupeneus multifasciatus*			Grenadier, Longarm	*Coelorinchus macrochir*		
Goatfish, Mexican	*Mulloidichthys dentatus*			Grenadier, Pacific	*Coryphaenoides acrolepis*		
Goatfish, Pfleuger's	*Mulloidichthys pfluegeri*			Grenadier, Roundnose	*Coryphaenoides rupestris*		
Goatfish, Pointed	*Parupeneus biaculeatus*			Groper, Eastern Blue	*Achoerodus viridis*		
Goatfish, Rosy	*Parupeneus rubescens*			Grouper, Areolate	*Epinephelus areolatus*		
Goatfish, Whitesaddle	*Parupeneus ciliatus*			Grouper, Black	*Mycteroperca bonaci*		
Goatfish, Yellowfin	*Mulloidichthys vanicolensis*						

Grouper, Blacksaddle	_Epinephelus Howlandi_			Grouper, Mottled	_Mycteroperca rubra_		
Grouper, Blue And Yellow	_Epinephelus flavocaeruleus_			Grouper, Moustache	_Epinephelus chabaudi_		
Grouper, Broomtail	_Mycteroperca xenarcha_			Grouper, Nassau	_Epinephelus striatus_		
Grouper, Brown-marbled	_Epinephelus fuscoguttatus_			Grouper, Netfin	_Epinephelus miliaris_		
Grouper, Brownspotted	_Epinephelus chlorostigma_			Grouper, Oblique-banded	_Epinephelus radiatus_		
Grouper, Camouflage	_Epinephelus polyphekadion_			Grouper, Olive	_Epinephelus cifuentesi_		
Grouper, Cloudy	_Epinephelus erythrurus_			Grouper, Orange-spotted	_Epinephelus coioides_		
Grouper, Comb	_Mycteroperca acutirostris_			Grouper, Potato	_Epinephelus tukula_		
Grouper, Comet	_Epinephelus morrhua_			Grouper, Red	_Epinephelus morio_		
Grouper, Convict	_Epinephelus septemfasciatus_			Grouper, Redmouth	_Aethaloperca rogaa_		
Grouper, Dot-dash	_Epinephelus poecilonotus_			Grouper, Red-tipped	_Epinephelus retouti_		
Grouper, Dusky	_Epinephelus marginatus_			Grouper, Sawtail	_Mycteroperca prionura_		
Grouper, Duskytail	_Epinephelus bleekeri_			Grouper, Small Scaled	_Epinephelus polylepis_		
Grouper, Gag	_Mycteroperca microlepis_			Grouper, Snowy	_Hyporthodus niveatus_		
Grouper, Giant	_Epinephelus lanceolatus_			Grouper, Snubnose	_Epinephelus macrospilos_		
Grouper, Goldblotch	_Epinephelus costae_			Grouper, Speckled Blue	_Epinephelus cyanopodus_		
Grouper, Goliath	_Epinephelus itajara_			Grouper, Spotted	_Epinephelus analogus_		
Grouper, Gulf	_Mycteroperca jordani_			Grouper, Starry	_Epinephelus labriformis_		
Grouper, Halfmoon	_Epinephelus rivulatus_			Grouper, Star-studded	_Hyporthodus niphobles_		
Grouper, Hawaiian	_Hyporthodus quernus_			Grouper, Striped	_Epinephelus latifasciatus_		
Grouper, Highfin	_Epinephelus maculatus_			Grouper, Striped Fin	_Epinephelus posteli_		
Grouper, Hong Kong	_Epinephelus akaara_			Grouper, Tiger	_Mycteroperca tigris_		
Grouper, Island	_Mycteroperca fusca_			Grouper, Warsaw	_Hyporthodus nigritus_		
Grouper, Leopard	_Mycteroperca rosacea_			Grouper, White	_Epinephelus aeneus_		
Grouper, Longfin	_Epinephelus quoyanus_			Grouper, White-blotched	_Epinephelus multinotatus_		
Grouper, Longspine	_Epinephelus longispinis_			Grouper, Yellow	_Epinephelus awoara_		
Grouper, Longtooth	_Epinephelus bruneus_			Grouper, Yellowedge	_Hyporthodus flavolimbatus_		
Grouper, Malabar	_Epinephelus malabaricus_			Grouper, Yellowfin	_Mycteroperca venenosa_		
Grouper, Marbled	_Dermatolepis inermis_			Grouper, Yellowmouth	_Mycteroperca interstitialis_		

Grouper, Yellowspotted	*Epinephelus timorensis*		
Grunt, African Striped	*Parapristipoma octolineatum*		
Grunt, Bastard	*Pomadasys incisus*		
Grunt, Biglip	*Plectorhinchus macrolepis*		
Grunt, Bluestriped	*Haemulon sciurus*		
Grunt, Burrito	*Anisotremus interruptus*		
Grunt, Burro	*Pomadasys crocro*		
Grunt, Greybar	*Haemulon sexfasciatum*		
Grunt, Longspine	*Pomadasys macracanthus*		
Grunt, Pacific Roncador	*Pomadasys bayanus*		
Grunt, Rubberlip	*Plectorhinchus mediterraneus*		
Grunt, Silver	*Pomadasys argenteus*		
Grunt, Sompant	*Pomadasys jubelini*		
Grunt, Spanish	*Haemulon macrostomum*		
Grunt, Tomtate	*Haemulon aurolineatum*		
Grunt, White	*Haemulon plumierii*		
Grunt, White	*Haemulopsis leuciscus*		
Grunter, Javelin	*Pomadasys kaakan*		
Grunter, Saddle	*Pomadasys maculatus*		
Guapote	*Parachromis dovii*		
Guapote, Jaguar	*Parachromis managuensis*		
Gudgeon, Northern Mud	*Ophiocara porocephala*		
Guitarfish, Banded	*Zapteryx exasperata*		
Guitarfish, Blackchin	*Rhinobatos cemiculus*		
Guitarfish, Giant	*Rhynchobatus djiddensis*		
Guitarfish, Lesser	*Zapteryx brevirostris*		
Guitarfish, Shovelnose	*Rhinobatos productus*		
Guitarfish, Thornback	*Platyrhinoidis triseriata*		
Guitarfish, Yellow	*Rhinobatos schlegelii*		

Gurnard, Bluefin	*Chelidonichthys kumu*		
Gurnard, Flying	*Dactylopterus volitans*		
Gurnard, Grey	*Eutrigla gurnardus*		
Gurnard, Red	*Chelidonichthys cuculus*		
Gurnard, Red	*Chelidonichthys spinosus*		
Gurnard, Spotted	*Pterygotrigla andertoni*		
Gurnard, Tub	*Chelidonichthys lucerna*		
Haddock	*Melanogrammus aeglefinus*		
Hairtail, Largehead	*Trichiurus lepturus*		
Hake, Argentine	*Merluccius hubbsi*		
Hake, Carolina	*Urophycis earllii*		
Hake, European	*Merluccius merluccius*		
Hake, Gulf	*Urophycis cirrata*		
Hake, Pacific	*Merluccius productus*		
Hake, Red	*Urophycis chuss*		
Hake, Silver	*Merluccius bilinearis*		
Hake, White	*Urophycis tenuis*		
Halfmoon	*Medialuna californiensis*		
Halibut, Atlantic	*Hippoglossus hippoglossus*		
Halibut, California	*Paralichthys californicus*		
Halibut, Greenland	*Reinhardtius hippoglossoides*		
Halibut, Pacific	*Hippoglossus stenolepis*		
Halibut, Shotted	*Eopsetta grigorjewi*		
Happy, Pink	*Sargochromis giardi*		
Hapuku	*Polyprion oxygeneios*		
Hardhead	*Mylopharodon conocephalus*		
Hawkfish, Giant	*Cirrhitus rivulatus*		
Herring, Atlantic	*Clupea harengus*		
Herring, Skipjack	*Alosa chrysochloris*		
Hind, Coral	*Cephalopholis miniata*		

Hind, Garish	*Cephalopholis igarashiensis*		
Hind, Golden	*Cephalopholis aurantia*		
Hind, Peacock	*Cephalopholis argus*		
Hind, Red	*Epinephelus guttatus*		
Hind, Rock	*Epinephelus adscensionis*		
Hind, Speckled	*Epinephelus drummondhayi*		
Hind, Tomato	*Cephalopholis sonnerati*		
Hogfish	*Lachnolaimus maximus*		
Hogfish, Golden-spot	*Bodianus perditio*		
Hogfish, Hawaiian	*Bodianus albotaeniatus*		
Hogfish, Spanish	*Bodianus rufus*		
Hogfish, Spotfin	*Bodianus pulchellus*		
Hogfish, Tarry	*Bodianus bilunulatus*		
Hokke	*Pleurogrammus azonus*		
Horsehead, White	*Branchiostegus albus*		
Horsehead, Yellow	*Branchiostegus auratus*		
Hottentot	*Pachymetopon blochii*		
Houndfish	*Tylosurus crocodilus*		
Houndfish, Mexican	*Tylosurus crocodilus fodiator*		
Houndfish, Red Sea	*Tylosurus choram*		
Huchen	*Hucho hucho*		
Huchen, Japanese	*Parahucho perryi*		
Ide	*Leuciscus idus*		
Inconnu	*Stenodus leucichthys*		
Isaki	*Parapristipoma trilineatum*		
Itoyoridai	*Nemipterus virgatus*		
Iwatoko-namazu	*Silurus lithophilus*		
Izukasago	*Scorpaena izensis*		
Jack, Almaco	*Seriola rivoliana*		
Jack, Bar	*Caranx ruber*		

Jack, Black	*Caranx lugubris*		
Jack, Bluntnose	*Hemicaranx amblyrhynchus*		
Jack, Cornish	*Mormyrops anguilloides*		
Jack, Cottonmouth	*Uraspis secunda*		
Jack, Crevalle	*Caranx hippos*		
Jack, Fortune	*Seriola peruana*		
Jack, Green	*Caranx caballus*		
Jack, Horse-eye	*Caranx latus*		
Jack, Longfin Crevalle	*Caranx fischeri*		
Jack, Pacific Crevalle	*Caranx caninus*		
Jack, Senegal	*Caranx senegallus*		
Jack, Whitetongue	*Uraspis helvola*		
Jack, Yellow	*Carangoides bartholomaei*		
Jacopever, False	*Sebastes capensis*		
Jacunda	*Crenicichla lugubris*		
Jacunda, Johans	*Crenicichla johanna*		
Jacunda, Lenticulated	*Crenicichla lenticulata*		
Jacunda, Marbled	*Crenicichla marmorata*		
Jacunda, Striped	*Crenicichla cincta*		
Jandia	*Rhamdia sebae*		
Jau	*Zungaro zungaro*		
Jawfish, Finespotted	*Opistognathus punctatus*		
Jawfish, Giant	*Opistognathus rhomaleus*		
Jobfish, Crimson	*Pristipomoides filamentosus*		
Jobfish, Golden Eye	*Pristipomoides flavipinnis*		
Jobfish, Goldflag	*Pristipomoides auricilla*		
Jobfish, Green	*Aprion virescens*		
Jobfish, Lavender	*Pristipomoides sieboldii*		
Jobfish, Rusty	*Aphareus rutilans*		
Jobfish, Small Toothed	*Aphareus furca*		
John Dory	*Zeus faber*		
Jumprock, Greater	*Moxostoma lachneri*		
Jundia	*Leiarius marmoratus*		

Jurupoca	*Hemisorubim platyrhynchos*			Largemouth, Thinface	*Serranochromis angusticeps*		
Kahawai	*Arripis trutta*			Lates, Forktail	*Lates microlepis*		
Kawakawa	*Euthynnus affinis*			Lates, Japanese	*Lates japonicus*		
Kelpfish, False	*Sebastiscus marmoratus*			Lau-lau	*Brachyplatystoma filamentosum*		
Kingfish, Butterfly	*Gasterochisma melampus*			Leaffish, Malayan	*Pristolepis fasciata*		
Kingfish, Gulf	*Menticirrhus littoralis*			Leatherjack, Longjaw	*Oligoplites altus*		
Kingfish, Northern	*Menticirrhus saxatilis*			Leatherjacket, Six-spined	*Meuschenia freycineti*		
Kingfish, Southern	*Menticirrhus americanus*			Leerfish	*Lichia amia*		
Kitsunedai	*Bodianus oxycephalus*			Lenok	*Brachymystax lenok*		
Kitsune-mebaru	*Sebastes vulpes*			Lenok, Stupid	*Brachymystax savinovi*		
Knifefish, Clown	*Chitala chitala*			Ling, Blue	*Molva dypterygia*		
Knifejaw, Natal	*Oplegnathus robinsoni*			Ling, European	*Molva molva*		
Kob	*Argyrosomus hololepidotus*			Lingcod	*Ophiodon elongatus*		
Kobudai	*Semicossyphus reticulatus*			Lionfish, Red	*Pterois volitans*		
Koheru	*Decapterus koheru*			Lizardfish, Atlantic	*Synodus saurus*		
Kokanee	*Oncorhynchus nerka*			Lizardfish, Bluntnose	*Trachinocephalus trachinus*		
Kokuni	*Chrysichthys cranchii*			Lizardfish, Inshore	*Synodus foetens*		
Korai-kitsunemebaru	*Sebastes ijimae*			Lizardfish, Lance	*Synodus sciuliceps*		
Kurobuchi-Tensu	*Iniistius geisha*			Lizardfish, Red	*Synodus ulae*		
Kurogarei	*Pseudopleuronectes obscurus*			Lizardfish, Slender	*Saurida elongata*		
Kuromenuke	*Sebastes glaucus*			Lizardfish, Wanieso	*Saurida wanieso*		
Kuro-sabafugu	*Lagocephalus gloveri*			Lookdown	*Selene vomer*		
Kurosoi	*Sebastes schlegelii*			Lord, Irish	*Hemilepidotus spinosus*		
Labeo Barbatulus	*Labeo barbatulus*			Lord, Red Irish	*Hemilepidotus hemilepidotus*		
Labeo, Orangefin	*Labeo calbasu*			Lord, Yellow Irish	*Hemilepidotus jordani*		
Ladyfish	*Elops saurus*			Lyretail, White-edged	*Variola albimarginata*		
Ladyfish, Hawaiian	*Elops hawaiensis*			Lyretail, Yellow-edged	*Variola louti*		
Ladyfish, Pacific	*Elops affinis*			Ma-Anago	*Conger myriaster*		
Ladyfish, Senegalese	*Elops senegalensis*			Machaca	*Brycon guatemalensis*		
Ladyfish, Springer	*Elops machnata*			Machaca (Sabalo Pipon)	*Brycon behreae*		
Largemouth, Humpback	*Serranochromis altus*			Mackerel, Atka	*Pleurogrammus monopterygius*		
Largemouth, Purple-faced	*Serranochromis macrocephalus*			Mackerel, Atlantic	*Scomber scombrus*		

Common Name	Scientific Name			Common Name	Scientific Name		
Mackerel, Atlantic Chub	Scomber colias			Manduba	Ageneiosus inermis		
Mackerel, Australian Spotted	Scomberomorus munroi			Maomao, Blue	Scorpis violacea		
Mackerel, Blue	Scomber australasicus			Maomao, Pink	Caprodon longimanus		
Mackerel, Broadbarred	Scomberomorus semifasciatus			Margate, Black	Anisotremus surinamensis		
Mackerel, Bullet	Auxis rochei			Margate, White	Haemulon album		
Mackerel, Cero	Scomberomorus regalis			Marimba	Diplodus argenteus		
Mackerel, Chub	Scomber japonicus			Marlin, Black	Istiompax indica		
Mackerel, Double-lined	Grammatorcynus bilineatus			Marlin, Blue	Makaira nigricans		
Mackerel, Frigate	Auxis thazard			Marlin, Blue	Makaira nigricans		
Mackerel, Greenback Horse	Trachurus declivis			Marlin, Striped	Kajikia audax		
Mackerel, Indian	Rastrelliger kanagurta			Marlin, White	Kajikia albida		
Mackerel, Indo-Pacific King	Scomberomorus guttatus			Matrincha	Brycon falcatus		
Mackerel, Japanese Jack	Trachurus japonicus			Meagre	Argyrosomus regius		
Mackerel, Japanese Spanish	Scomberomorus niphonius			Meagre, Japanese	Argyrosomus japonicus		
Mackerel, King	Scomberomorus cavalla			Mebaru	Sebastes inermis		
Mackerel, Monterey Spanish	Scomberomorus concolor			Megrim	Lepidorhombus whiffiagonis		
Mackerel, Narrowbarred	Scomberomorus commerson			Mejina	Girella punctata		
Mackerel, Pacific Jack	Trachurus symmetricus			Menada	Liza haematocheila		
Mackerel, Pacific Sierra	Scomberomorus sierra			Mihara-hanadai	Giganthias immaculatus		
Mackerel, Shark	Grammatorcynus bicarinatus			Milkfish	Chanos chanos		
Mackerel, Snake	Gempylus serpens			Moga	Tomocichla tuba		
Mackerel, Spanish	Scomberomorus maculatus			Mojarra, Striped	Eugerres plumieri		
Mackerel, West African Spanish	Scomberomorus tritor			Mojarra, Yellow	Caquetaia kraussii		
Mackerel, Yellowtail Horse	Trachurus novaezelandiae			Mojarra, Yellowfin	Gerres cinereus		
Madai	Pagrus major			Moncholo	Pimelodus albicans		
Mahseer	Tor tor			Moncholo, Amarillo	Megalonema platanum		
Mahseer, Deccan	Tor khudree			Monkfish, European	Squatina squatina		
Mahseer, Golden	Tor putitora			Mooneye	Hiodon tergisus		
Mahseer, Stracheyi	Neolissochilus stracheyi			Moray, Argus	Muraena argus		
Mahseer, Thai	Tor tambroides			Moray, Australian Mottled	Gymnothorax prionodon		
Mandarin Fish, Leopard	Siniperca scherzeri			Moray, Banded	Gymnothorax rueppelliae		

Moray, Blackedge	*Gymnothorax nigromarginatus*			Morwong, Peruvian	*Cheilodactylus variegatus*		
Moray, Blacktail	*Gymnothorax kolpos*			Morwong, Spottedtail	*Goniistius zonatus*		
Moray, California	*Gymnothorax mordax*			Mrigal	*Cirrhinus cirrhosus*		
Moray, Chestnut	*Gymnothorax castaneus*			Mullet, Diamond	*Planiliza alata*		
Moray, Geometric	*Gymnothorax griseus*			Mullet, Hog	*Joturus pichardi*		
Moray, Green	*Gymnothorax funebris*			Mullet, Liza	*Mugil liza*		
Moray, Grey	*Gymnothorax nubilus*			Mullet, Mountian	*Agonostomus monticola*		
Moray, Highfin	*Gymnothorax pseudothyrsoideus*			Mullet, Striped	*Mugil cephalus*		
				Mullet, Thicklip	*Chelon labrosus*		
Moray, Honeycomb	*Muraena melanotis*			Mullet, Thinlip	*Liza ramada*		
Moray, Hourglass	*Muraena clepsydra*			Mullet, White	*Mugil curema*		
Moray, Jewel	*Muraena lentiginosa*			Murasoi	*Sebastes pachycephalus*		
Moray, Laced	*Gymnothorax favagineus*			Muroranginpo	*Pholidapus dybowskii*		
Moray, Mediterranean	*Muraena helena*			Muskellunge	*Esox masquinongy*		
Moray, Peppered	*Gymnothorax pictus*						
Moray, Purplemouth	*Gymnothorax vicinus*			Muskellunge, Tiger	*Esox masquinongy x Esox lucius*		
Moray, Reticulate Hookjaw	*Enchelycore lichenosa*			Musselcracker, Black	*Cymatoceps nasutus*		
Moray, Slender Giant	*Strophidon sathete*			Myleus, Redhook	*Myloplus rubripinnis*		
Moray, Snowflake	*Echidna nebulosa*			Nagagaji	*Zoarces elongatus*		
Moray, Spotted	*Gymnothorax moringa*			Nase	*Chondrostoma nasus*		
Moray, Stout	*Gymnothorax eurostus*			Needlefish, Atlantic	*Strongylura marina*		
Moray, Turkey	*Gymnothorax meleagris*			Needlefish, Flat	*Ablennes hians*		
Moray, Undulated	*Gymnothorax undulatus*			Needlefish, Keel-jawed	*Tylosurus acus melanotus*		
Moray, Viper	*Enchelynassa canina*			Needlefish, Pacific Agujon	*Tylosurus pacificus*		
Moray, Whitemargin	*Gymnothorax albimarginatus*			Nembwe	*Serranochromis robustus*		
Moray, Yellow	*Gymnothorax prasinus*			Nigoi	*Hemibarbus barbus*		
Moray, Yellow-edged	*Gymnothorax flavimarginatus*			Niji-kajika	*Alcichthys elongatus*		
Moray, Zebra	*Gymnomuraena zebra*			Nkupe	*Distichodus mossambicus*		
Morwong, Blackbarred	*Goniistius quadricornis*			Oilfish	*Ruvettus pretiosus*		
				Okina-mejina	*Girella mezina*		
				Okin-buna	*Carassius auratus buergeri*		
				Okuchi-ishinagi	*Stereolepis doederleini*		

Oni-hige	*Coelorinchus gilberti*			Parrotfish, Midnight	*Scarus coelestinus*		
Oniokoze	*Inimicus japonicus*			Parrotfish, Pacific Longnose	*Hipposcarus longiceps*		
Opah	*Lampris incognitus*			Parrotfish, Rainbow	*Scarus guacamaia*		
Opaleye	*Girella nigricans*			Parrotfish, Redtail	*Sparisoma chrysopterum*		
Oscar	*Astronotus ocellatus*			Parrotfish, Rivulated	*Scarus rivulatus*		
Oshitabirame	*Cynoglossus bilineatus*			Parrotfish, Singapore	*Scarus prasiognathos*		
Ougon-murasoi	*Sebastes pachycephalus nudus*			Parrotfish, Steephead	*Chlorurus microrhinos*		
Pacu, Black	*Piaractus brachypomus*			Parrotfish, Stoplight	*Sparisoma viride*		
Pacu, Borracha	*Myleus pacu*			Parrotfish, Yellowband	*Scarus schlegeli*		
Pacu, Caranha	*Piaractus mesopotamicus*			Parrotfish, Yellow-tail	*Scarus hypselopterus*		
Pacu, Planquettei	*Myloplus planquettei*			Parrotperch, Japanese	*Oplegnathus fasciatus*		
Paloma	*Brycon rubricauda*			Parrotperch, Spotted	*Oplegnathus punctatus*		
Palometa	*Trachinotus goodei*			Pati	*Luciopimelodus pati*		
Pandora	*Pagellus erythrinus*			Peacock, Blackstriped	*Cichla intermedia*		
Pangasius, Giant	*Pangasius sanitwongsei*			Peacock, Blue	*Cichla piquiti*		
Pangasius, Shortbarbel	*Pangasius micronemus*			Peacock, Butterfly	*Cichla ocellaris*		
Pangasius, Spot	*Pangasius larnaudii*			Peacock, Jariina	*Cichla jariina*		
Parrotfish, Blue	*Scarus coeruleus*			Peacock, Melaniae	*chicla maleniae*		
Parrotfish, Blue-barred	*Scarus ghobban*			Peacock, Melaniae	*Cichla melaniae*		
Parrotfish, Bower's	*Chlorurus bowersi*			Peacock, Mirianae	*Cichla mirianae*		
Parrotfish, Chameleon	*Scarus chameleon*			Peacock, Orinoco	*Cichla orinocensis*		
Parrotfish, Common	*Scarus psittacus*			Peacock, Pinima	*Cichla pinima*		
Parrotfish, Daisy	*Chlorurus sordidus*			Peacock, Pleiozona	*Cichla pleiozona*		
Parrotfish, Darktail	*Scarus fuscocaudalis*			Peacock, Popoca	*Cichla monoculus*		
Parrotfish, Dusky	*Scarus niger*			Peacock, Speckled	*Cichla temensis*		
Parrotfish, Ember	*Scarus rubroviolaceus*			Peacock, Taua	*Cichla nigromaculata*		
Parrotfish, Festive	*Scarus festivus*			Peacock, Thyrorus	*Cichla thyrorus*		
Parrotfish, Forsten's	*Scarus forsteni*			Peacock, Vazzoleri	*Cichla vazzoleri*		
Parrotfish, Green Humphead	*Bolbometopon muricatum*			Peacock, Yellow	*Cichla kelberi*		
Parrotfish, Japanese	*Calotomus japonicus*			Pejerrey	*Odontesthes bonariensis*		
				Pellona, Amazon	*Pellona castelnaeana*		
				Pellona, Yellowfin River	*Pellona flavipinnis*		
				Perch, Chinese	*Siniperca chuatsi*		

Perch, Creole	*Percichthys trucha*			Pinook	*Oncorhynchus gorbuscha x O. tshawytscha*		
Perch, European	*Perca fluviatilis*			Piracatinga	*Calophysus macropterus*		
Perch, Nile	*Lates niloticus*			Pirambeba	*Serrasalmus humeralis*		
Perch, Pile	*Rhacochilus vacca*			Piramutaba	*Brachyplatystoma vaillantii*		
Perch, Red Gurnard	*Helicolenus percoides*			Piranha, Black	*Serrasalmus rhombeus*		
Perch, Sacramento	*Archoplites interruptus*			Piranha, Black Spot	*Pygocentrus cariba*		
Perch, Sunrise	*Caprodon schlegelii*			Piranha, Manueli's	*Serrasalmus manueli*		
Perch, White	*Morone americana*			Piranha, Red	*Pygocentrus nattereri*		
Perch, Yellow	*Perca flavescens*			Piranha, Serrulatus	*Serrasalmus serrulatus*		
Permit	*Trachinotus falcatus*			Piranha, White	*Serrasalmus spilopleura*		
Piabanha	*Brycon insignis*			Piraputanga	*Brycon hilarii*		
Piau	*Leporinus piau*			Pollack, European	*Pollachius pollachius*		
Piau, Red	*Leporinus brunneus*			Pollock	*Pollachius virens*		
Piavucu	*Leporinus macrocephalus*			Pollock, Alaska	*Theragra chalcogramma*		
Pickerel, Chain	*Esox niger*			Pomfret	*Taractes rubescens*		
Pickerel, Grass	*Esox americanus vermiculatus*			Pomfret, Bigscale	*Taractichthys longipinnis*		
Pickerel, Redfin	*Esox americanus americanus*			Pomfret, Brevort's	*Eumegistus brevorti*		
Picuda	*Salminus affinis*			Pomfret, Lustrous	*Eumegistus illustris*		
Pigfish	*Orthopristis chrysoptera*			Pomfret, Pacific	*Brama japonica*		
Pigfish, Red	*Bodianus unimaculatus*			Pompano, African	*Alectis ciliaris*		
Pigfish, Western	*Bodianus vulpinus*			Pompano, Blackblotch	*Trachinotus kennedyi*		
Pigfish, Yellowfin	*Bodianus flavipinnis*			Pompano, Florida	*Trachinotus carolinus*		
Pike, Amur	*Esox Reichertii*			Pompano, Gafftopsail	*Trachinotus rhodopus*		
Pike, Northern	*Esox lucius*			Pompano, Irish	*Diapterus auratus*		
Pike-characin, Golden	*Boulengerella lucius*			Pompano, Snubnose	*Trachinotus blochi*		
Pike-conger, Common	*Muraenesox bagio*			Pompano, Southern	*Trachinotus africanus*		
Pike-conger, Daggertooth	*Muraenesox cinereus*			Porae	*Nemadactylus douglasii*		
Pikeminnow, Northern	*Ptychocheilus oregonensis*			Porcupinefish, Black-blotched	*Diodon liturosus*		
Pikeminnow, Sacramento	*Ptychocheilus grandis*			Porcupinefish, Longspine	*Diodon holocanthus*		
Pinfish	*Lagodon rhomboides*			Porcupinefish, Spot-fin	*Diodon hystrix*		

Porgy, Black	*Acanthopagrus schlegelii*			Puffer, Whitespotted	*Arothron hispidus*		
Porgy, Grass	*Calamus arctifrons*			Puffer, Yellowfin	*Takifugu xanthopterus*		
Porgy, Jolthead	*Calamus bajonado*			Pufferfish, Japanese	*Takifugu rubripes*		
Porgy, Knobbed	*Calamus nodosus*			Pumpkinseed	*Lepomis gibbosus*		
Porgy, Littlehead	*Calamus proridens*			Queenfish, Doublespotted	*Scomberoides lysan*		
Porgy, Pacific	*Calamus brachysomus*			Queenfish, Needlescaled	*Scomberoides tol*		
Porgy, Pluma	*Calamus pennatula*			Queenfish, Talang	*Scomberoides commersonianus*		
Porgy, Red	*Pagrus pagrus*			Quillback	*Carpiodes cyprinus*		
Porgy, Saucereye	*Calamus calamus*			Ratfish, Spotted	*Hydrolagus colliei*		
Porgy, Sheepshead	*Calamus penna*			Raven, Sea	*Hemitripterus villosus*		
Porgy, Whitebone	*Calamus leucosteus*			Ray, Backwater Butterfly	*Gymnura natalensis*		
Porkfish	*Anisotremus virginicus*			Ray, Bat	*Myliobatis californica*		
Pouting	*Trisopterus luscus*			Ray, Blonde	*Raja brachyura*		
Powan	*Coregonus lavaretus*			Ray, Bluespotted Ribbontail	*Taeniura lymma*		
Prickleback, Monkeyface	*Cebidichthys violaceus*			Ray, Bull	*Aetomylaeus bovinus*		
Prochilod, Streaked	*Prochilodus lineatus*			Ray, Bull	*Myliobatis australis*		
Prowfish	*Zaprora silenus*			Ray, California Butterfly	*Gymnura marmorata*		
Puddingwife	*Halichoeres radiatus*			Ray, Common Eagle	*Myliobatis aquila*		
Puffer, Blunthead	*Sphoeroides pachygaster*			Ray, Discus	*Paratrygon aiereba*		
Puffer, Bullseye	*Sphoeroides annulatus*			Ray, Eagle	*Myliobatis tenuicaudatus*		
Puffer, Cheeseman's	*Lagocephalus cheesemanii*			Ray, Japanese Butterfly	*Gymnura japonica*		
Puffer, Guineafowl	*Arothron meleagris*			Ray, Japanese Eagle	*Myliobatis tobijei*		
Puffer, Longnose	*Sphoeroides lobatus*			Ray, Longheaded Eagle	*Aetobatus flagellum*		
Puffer, Northern	*Sphoeroides maculatus*			Ray, Painted	*Raja microocellata*		
Puffer, Oceanic	*Lagocephalus lagocephalus*			Ray, Pale	*Rajella lintea*		
Puffer, Panther	*Takifugu pardalis*			Ray, Round Ribbontail	*Taeniurops meyeni*		
Puffer, Prickly	*Ephippion guttifer*			Ray, Southern Fiddler	*Trygonorrhina fasciata*		
Puffer, Purple	*Takifugu porphyreus*			Ray, Spiny Butterfly	*Gymnura altavela*		
Puffer, Purple	*Takifugu vermicularis*			Ray, Spotted	*Raja montagui*		
Puffer, Smooth	*Lagocephalus laevigatus*			Ray, Spotted Eagle	*Aetobatus narinari*		

Ray, Starry	*Raja asterias*		
Ray, Thornback	*Raja clavata*		
Ray, Undulate	*Raja undulata*		
Razorfish, Peacock	*Iniistius pavo valenciennes*		
Rebeca	*Megalodoras uranoscopus*		
Red Devil	*Amphilophus labiatus*		
Redcoat	*Sargocentron rubrum*		
Redfin, Pacific	*Tribolodon brandti*		
Redfish	*Centroberyx affinis*		
Redfish, Acadian	*Sebastes fasciatus*		
Redfish, Golden	*Sebastes norvegicus*		
Redhorse, Black	*Moxostoma duquesnei*		
Redhorse, Blacktail	*Moxostoma poecilurum*		
Redhorse, Golden	*Moxostoma erythrurum*		
Redhorse, Gray	*Moxostoma congestum*		
Redhorse, Greater	*Moxostoma valenciennesi*		
Redhorse, Notchlip	*Moxostoma collapsum*		
Redhorse, Pealip	*Moxostoma pisolabrum*		
Redhorse, River	*Moxostoma carinatum*		
Redhorse, Shorthead	*Moxostoma macrolepidotum*		
Redhorse, Silver	*Moxostoma anisurum*		
Redhorse, V-lip	*Moxostoma pappillosum*		
Redtail, Surfperch	*Amphistichus rhodoterus*		
Remora, Common	*Remora remora*		
Ringwrasse, Pastel	*Hologymnosus doliatus*		
Roach	*Rutilus rutilus*		
Roach, Danube	*Rutilus pigus*		
Roach-Bream	*Rutillus rutilus x Abramis brama*		
Rock-bacu	*Lithodoras dorsalis*		
Rockcod, Bluelined	*Cephalopholis formosa*		
Rockfish, Bank	*Sebastes rufus*		

Rockfish, Black	*Sebastes melanops*		
Rockfish, Black-and-yellow	*Sebastes chrysomelas*		
Rockfish, Blackgill	*Sebastes melanostomus*		
Rockfish, Blue	*Sebastes mystinus*		
Rockfish, Bronzespotted	*Sebastes gilli*		
Rockfish, Brown	*Sebastes auriculatus*		
Rockfish, Canary	*Sebastes pinniger*		
Rockfish, Chameleon	*Sebastes phillipsi*		
Rockfish, China	*Sebastes nebulosus*		
Rockfish, Copper	*Sebastes caurinus*		
Rockfish, Deacon	*Sebastes diaconus*		
Rockfish, Dusky	*Sebastes ciliatus*		
Rockfish, Flag	*Sebastes rubrivinctus*		
Rockfish, Goldeye	*Sebastes thompsoni*		
Rockfish, Gopher	*Sebastes carnatus*		
Rockfish, Grass	*Sebastes rastrelliger*		
Rockfish, Greenspotted	*Sebastes chlorostictus*		
Rockfish, Greenstriped	*Sebastes elongatus*		
Rockfish, Honeycomb	*Sebastes umbrosus*		
Rockfish, Kelp	*Sebastes atrovirens*		
Rockfish, Mexican	*Sebastes macdonaldi*		
Rockfish, Olive	*Sebastes serranoides*		
Rockfish, Pink	*Sebastes eos*		
Rockfish, Quillback	*Sebastes maliger*		
Rockfish, Redbanded	*Sebastes babcocki*		
Rockfish, Redstripe	*Sebastes proriger*		
Rockfish, Rosethorn	*Sebastes helvomaculatus*		
Rockfish, Rougheye	*Sebastes aleutianus*		
Rockfish, Shortraker	*Sebastes borealis*		
Rockfish, Silvergray	*Sebastes brevispinis*		

Rockfish, Speckled	*Sebastes ovalis*			Salmon, Chinook-coho	*Oncorhynchus tshawytscha x O. kisutch*		
Rockfish, Splitnose	*Sebastes diploproa*			Salmon, Chum	*Oncorhynchus keta*		
Rockfish, Starry	*Sebastes constellatus*			Salmon, Coho	*Oncorhynchus kisutch*		
Rockfish, Sunset	*Sebastes crocotulus*			Salmon, Pink	*Oncorhynchus gorbuscha*		
Rockfish, Tiger	*Sebastes nigrocinctus*			Salmon, Sockeye	*Oncorhynchus nerka*		
Rockfish, Vermillion	*Sebastes miniatus*			Samson Fish	*Seriola hippos*		
Rockfish, Widow	*Sebastes entomelas*			Sand Diver	*Synodus intermedius*		
Rockfish, Yelloweye	*Sebastes ruberrimus*			Sandbass, Parrot	*Paralabrax loro*		
Rockfish, Yellowtail	*Sebastes flavidus*			Sanddab, Pacific	*Citharichthys sordidus*		
Rohu	*Labeo rohita*			Sandperch, Argentinian	*Pseudopercis semifasciata*		
Rohu, Kali	*Labeo oycheilus*			Sandperch, Brazilian	*Pinguipes brasilianus*		
Roosterfish	*Nematistius pectoralis*			Sandperch, Greater	*Diplectrum maximum*		
Rosefish, Blackbelly	*Helicolenus dactylopterus*			Sandperch, Namorado	*Pseudopercis numida*		
Rubyfish	*Plagiogeneion rubiginosum*			Sankomenuke	*Sebastes flammeus*		
Rubyfish, Japanese	*Erythrocles schlegelii*			Sardinata	*Brycon whitei*		
Rudd	*Scardinius erythrophthalmus*			Sargo	*Anisotremus davidsonii*		
Rudderfish	*Centrolophus niger*			Sauger	*Sander canadense*		
Rudderfish, Banded	*Seriola zonata*			Saugeye	*Stizostedion vitreum x S. canadense*		
Runner, Blue	*Caranx crysos*						
Runner, Rainbow	*Elagatis bipinnulata*			Sawshark, Japanese	*Pristiophorus japonicus*		
Sabaleta	*Brycon henni*			Sawtail, Scalpel	*Prionurus scalprum*		
Sabalo	*Brycon melanopterus*			Scabbardfish, Channel	*Evoxymetopon taeniatus*		
Sablefish	*Anoplopoma fimbria*			Scabbardfish, Silver	*Lepidopus caudatus*		
Sackfish, American	*Neoepinnula americana*			Scad, Amberstripe	*Decapterus muroadsi*		
Sailfish, Atlantic	*Istiophorus platypterus*			Scad, Bigeye	*Selar crumenophthalmus*		
Sailfish, Pacific	*Istiophorus platypterus*			Scad, False	*Caranx rhonchus*		
Sailors Choice	*Haemulon parra*			Scad, Mackerel	*Decapterus macarellus*		
Salema	*Sarpa salpa*			Scad, Oxeye	*Selar boops*		
Salmon, Atlantic	*Salmo salar*			Scad, Shortfin	*Decapterus macrosoma*		
Salmon, Atlantic (landlocked)	*Salmo salar*			Scad, Torpedo	*Megalaspis cordyla*		
Salmon, Chinook	*Oncorhynchus tshawytscha*						

Scamp	*Mycteroperca phenax*		
Scombrops, Atlantic	*Scombrops oculatus*		
Scorpionfish, Black	*Scorpaena porcus*		
Scorpionfish, California	*Scorpaena guttata*		
Scorpionfish, Darkblotch	*Scorpaena histrio*		
Scorpionfish, Large-headed	*Pontinus macrocephalus*		
Scorpionfish, Orange	*Scorpaena scrofa*		
Scorpionfish, Peruvian	*Scorpaena afuerae*		
Scorpionfish, Red	*Pontinus furcirhinus*		
Scorpionfish, Spinycheek	*Neomerinthe hemingwayi*		
Scorpionfish, Spotback	*Pontinus vaughani*		
Scorpionfish, Spotted	*Scorpaena plumieri*		
Scorpionfish, Stone	*Scorpaena mystes*		
Sculpin, Antlered	*Enophrys diceraus*		
Sculpin, Buffalo	*Enophrys Bison*		
Sculpin, Great	*Myoxocephalus polyacanthocephalus*		
Sculpin, Plain	*Myoxocephalus jaok*		
Sculpin, Shorthorn	*Myoxocephalus scorpius*		
Sculpin, Spinyhead	*Dasycottus setiger*		
Sculpin, Stellar's	*Myoxocephalus stelleri*		
Scup	*Stenotomus chrysops*		
Seabass, Argentine	*Acanthistius brasilianus*		
Seabass, Blackfin	*Lateolabrax latus*		
Seabass, Japanese	*Lateolabrax japonicus*		
Seabass, Peruvian Rock	*Paralabrax humeralis*		
Seabass, Rosy	*Doederleinia berycoides*		
Seabass, Spotted	*Dicentrarchus punctatus*		
Seabass, White	*Atractoscion nobilis*		
Seabream, Arabian Yellowfin	*Acanthopagrus arabicus*		

Seabream, Axillary	*Pagellus acarne*		
Seabream, Black	*Spondyliosoma cantharus*		
Seabream, Bluespotted	*Pagrus caeruleostictus*		
Seabream, Crimson	*Dentex tumifrons*		
Seabream, Daggerhead	*Chrysoblephus cristiceps*		
Seabream, Gilthead	*Sparus aurata*		
Seabream, Goldlined	*Rhabdosargus sarba*		
Seabream, Japanese Blue-spotted	*Amamiichthys matsubarai*		
Seabream, Moroccan White	*Diplodus sargus cadenati*		
Seabream, New Large-eye	*Gymnocranius obesus*		
Seabream, Okinawa	*Acanthopagrus sivicolus*		
Seabream, Red Stumpnose	*Chrysoblephus gibbiceps*		
Seabream, Redbanded	*Pagrus auriga*		
Seabream, Saddle	*Oblada melanura*		
Seabream, Scotsman	*Polysteganus praeorbitalis*		
Seabream, Sharpsnout	*Diplodus puntazzo*		
Seabream, Spottail	*Diplodus holbrookii*		
Seabream, Striped	*Lithognathus mormyrus*		
Seabream, White	*Diplodus sargus*		
Seabream, Yellowback	*Dentex tumifrons*		
Seabream, Yellowfin	*Acanthopagrus latus*		
Seabream, Zebra	*Diplodus cervinus*		
Seaperch, Spotted Scale	*Lutjanus johnii*		
Seaperch, Striped	*Embiotoca lateralis*		
Searobin, Striped	*Prionotus evolans*		
Seatrout, Sand	*Cynoscion arenarius*		
Seatrout, Silver	*Cynoscion nothus*		
Seatrout, Spotted	*Cynoscion nebulosus*		
Seerfish, Chinese	*Scomberomorus sinensis*		
Seerfish, Kanadi	*Scomberomorus plurilineatus*		

Sennet, Southern	Sphyraena picudilla			Shark, Greenland	Somniosus microcephalus		
Sergeant Baker	Latropiscis purpurissatus			Shark, Grey Bamboo	Chiloscyllium griseum		
Serrana, Blacktail	Serranus atricauda			Shark, Gulper	Centrophorus granulosus		
Seventy-four	Polysteganus undulosus			Shark, Gulper	Centrophorus uyato		
Shad, Allis	Alosa alosa			Shark, Gummy	Mustelus antarcticus		
Shad, American	Alosa sapidissima			Shark, Horn	Heterodontus francisci		
Shad, Gizzard	Dorosoma cepedianum			Shark, Japanese Bullhead	Heterodontus japonicus		
Shad, Hickory	Alosa mediocris			Shark, Japanese Tope	Hemitriakis japanica		
Shad, Twaite	Alosa fallax			Shark, Lemon	Negaprion brevirostris		
Shark, Atlantic Sharpnose	Rhizoprionodon terraenovae			Shark, Leopard	Triakis semifasciata		
Shark, Banded	Triakis scyllium			Shark, Milk	Rhizoprionodon acutus		
Shark, Bigeye Thresher	Alopias superciliosus			Shark, Mosaic Gulper	Centrophorus tessellatus		
Shark, Bignose	Carcharhinus altimus			Shark, Narrowtooth	Carcharhinus brachyurus		
Shark, Blackmouth Cat	Galeus melastomus			Shark, Night	Carcharhinus signatus		
Shark, Blacknose	Carcharhinus acronotus			Shark, Nurse	Ginglymostoma cirratum		
Shark, Blacktail	Carcharhinus amblyrhynchos			Shark, Oceanic Whitetip	Carcharhinus longimanus		
Shark, Blacktip	Carcharhinus limbatus			Shark, Pig-eye	Carcharhinus amboinensis		
Shark, Blacktip Reef	Carcharhinus melanopterus			Shark, Porbeagle	Lamna nasus		
Shark, Blotchy Swell	Cephaloscyllium umbratile			Shark, Port Jackson	Heterodontus portusjacksoni		
Shark, Blue	Prionace glauca			Shark, Salmon	Lamna ditropis		
Shark, Bonnethead	Sphyrna tiburo			Shark, Sand Tiger	Carcharias taurus		
Shark, Brown Smooth-hound	Mustelus henlei			Shark, Sandbar	Carcharhinus plumbeus		
Shark, Bull	Carcharhinus leucas			Shark, Scalloped Hammerhead	Sphyrna lewini		
Shark, Caribbean Reef	Carcharhinus perezii			Shark, Sevengill	Notorynchus cepedianus		
Shark, Dusky	Carcharhinus obscurus			Shark, Sharpnose Sevengill	Heptranchias perlo		
Shark, Dwarf Gulper	Centrophorus atromarginatus			Shark, Finetooth	Carcharhinus isodon			Shark, Shortfin Mako	Isurus oxyrinchus		
Shark, Finetooth	Carcharhinus isodon			Shark, Sicklefin Lemon	Negaprion acutidens		
Shark, Frog	Somniosus longus			Shark, Silky	Carcharhinus falciformis		
Shark, Galapagos	Carcharhinus galapagensis			Shark, Silvertip	Carcharhinus albimarginatus		
Shark, Ghost	Callorhinchus milii			Shark, Sixgilled	Hexanchus griseus		
Shark, Great Hammerhead	Sphyrna mokarran						

Shark, Smallfin Gulper	*Centrophorus moluccensis*				Skate, Clearnose	*Raja eglanteria*		
Shark, Smooth Hammerhead	*Sphyrna zygaena*				Skate, Little	*Leucoraja erinacea*		
Shark, Spinner	*Carcharhinus brevipinna*				Skate, Starry	*Amblyraja radiata*		
Shark, Spot-tail	*Carcharhinus sorrah*				Skipjack, Black	*Euthynnus lineatus*		
Shark, Thresher	*Alopias vulpinus*				Sleeper, Bigmouth	*Gobiomorus dormitor*		
Shark, Tiger	*Galeocerdo cuvier*				Slickhead, Lloyd's	*Narcetes lloydi*		
Shark, Tope	*Galeorhinus galeus*				Slimehead, Darwin's	*Gephyroberyx Darwinii*		
Shark, Velvet Belly Lantern	*Etmopterus spinax*				Smoothhound, Florida	*Mustelus norrisi*		
Shark, White	*Carcharodon carcharias*				Smoothhound, Gray	*Mustelus californicus*		
Shark, Whitetip Reef	*Triaenodon obesus*				Smooth-hound, Narrownose	*Mustelus schmitti*		
Sharkminnow, Black	*Labeo chrysophekadion*				Smoothhound, Spotted Estuary	*Mustelus lenticulatus*		
Sharkminnow, Tricolor	*Balantiocheilos melanopterus*				Smoothhound, Starry	*Mustelus asterias*		
Sharksucker	*Echeneis naucrates*				Smoothhound, Star-spotted	*Mustelus manazo*		
Sharksucker, Whitefin	*Echeneis neucratoides*				Snakehead, Blotched	*Channa maculata*		
Sheatfish, Whisker	*Phalacronotus bleekeri*				Snakehead, Chevron	*Channa striata*		
Sheephead, California	*Semicossyphus pulcher*				Snakehead, Emperor	*Channa marulioides*		
Sheepshead	*Archosargus probatocephalus*				Snakehead, Giant	*Channa micropeltes*		
Shimazoi	*Sebastes trivittatus*				Snakehead, Great	*Channa marulius*		
Shimofurikajika	*Myoxocephalus brandtii*				Snakehead, Northern	*Channa argus*		
Shiner, Golden	*Notemigonus crysoleucas*				Snakehead, Splendid	*Channa lucius*		
Shiner, Silver	*Notropis photogenis*				Snapper	*Pagrus auratus*		
Shosai-fugu	*Takifugu snyderi*				Snapper, African Brown	*Lutjanus dentatus*		
Sicklefish	*Drepane punctata*				Snapper, African Red	*Lutjanus agennes*		
Sierra, Atlantic	*Scomberomorus brasiliensis*				Snapper, Atlantic Cubera	*Lutjanus cyanopterus*		
Silver Eye	*Polymixia japonica*				Snapper, Black	*Apsilus dentatus*		
Silver-Biddy, Strongspine	*Gerres Longirostris*				Snapper, Black And White	*Macolor niger*		
Silverside, Jack	*Atherinopsis californiensis*				Snapper, Blackfin	*Lutjanus buccanella*		
Skate	*Raja batis*				Snapper, Blacktail	*Lutjanus fulvus*		
Skate, Big	*Raja binoculata*				Snapper, Brown-striped Red	*Lutjanus vitta*		
Skate, California	*Raja inornata*				Snapper, Colorado	*Lutjanus colorado*		
					Snapper, Common Blueline	*Lutjanus kasmira*		
					Snapper, Dog	*Lutjanus jocu*		
					Snapper, Emperor	*Lutjanus sebae*		

Snapper, Flame	*Etelis coruscans*			Snapper, Vermillion	*Rhomboplites aurorubens*	
Snapper, Freshwater	*Lutjanus fuscescens*			Snapper, Yellow	*Lutjanus argentiventris*	
Snapper, Gorean	*Lutjanus goreensis*			Snapper, Yellow-lined	*Lutjanus rufolineatus*	
Snapper, Greenbar	*Hoplopagrus guentherii*			Snapper, Yellowstreaked	*Lutjanus lemniscatus*	
Snapper, Grey	*Lutjanus griseus*			Snapper, Yellowtail	*Ocyurus chrysurus*	
Snapper, Humpback Red	*Lutjanus gibbus*			Snoek	*Thyrsites atun*	
Snapper, Japanese	*Paracaesio caerulea*			Snoek, Black	*Thyrsitoides marleyi*	
Snapper, Lane	*Lutjanus synagris*			Snook, Common	*Centropomus undecimalis*	
Snapper, Mahogany	*Lutjanus mahogoni*			Snook, Fat	*Centropomus parallelus*	
Snapper, Malabar	*Lutjanus malabaricus*			Snook, Mexican	*Centropomus poeyi*	
Snapper, Mangrove Red	*Lutjanus argentimaculatus*			Snook, Pacific Black	*Centropomus nigrescens*	
Snapper, Mullet	*Lutjanus aratus*			Snook, Pacific Blackfin	*Centropomus medius*	
Snapper, Mutton	*Lutjanus analis*			Snook, Pacific White	*Centropomus viridis*	
Snapper, Oblique-banded	*Pristipomoides zonatus*			Snook, Swordspine	*Centropomus ensiferus*	
Snapper, One-spot	*Lutjanus monostigma*			Snook, Tarpon	*Centropomus pectinatus*	
Snapper, Pacific Cubera	*Lutjanus novemfasciatus*			Sohachi	*Cleisthenes pinetorum*	
Snapper, Pacific Red	*Lutjanus peru*			Soldierfish, Bigscale	*Myripristis berndti*	
Snapper, Papuan Black	*Lutjanus goldiei*			Soldierfish, Blacktip	*Myripristis botche*	
Snapper, Queen	*Etelis oculatus*			Soldierfish, Japanese	*Ostichthys japonicus*	
Snapper, Red	*Lutjanus campechanus*			Sole	*Solea solea*	
Snapper, Ruby	*Etelis carbunculus*			Sole, Dusky	*Lepidopsetta mochigarei*	
Snapper, Russell's	*Lutjanus russellii*			Sole, Fantail	*Xystreurys liolepis*	
Snapper, Saddle-back	*Paracaesio kusakarii*			Sole, Lemon	*Microstomus kitt*	
Snapper, Scarlet	*Etelis radiosus*			Sole, Northern	*Lepidopsetta polyxystra*	
Snapper, Schoolmaster	*Lutjanus apodus*			Sole, Pacific Sand	*Psettichthys melanostictus*	
Snapper, Silk	*Lutjanus vivanus*			Sole, Southern Rock	*Lepidopsetta bilineata*	
Snapper, Spanish Flag	*Lutjanus carponotatus*			Sole, Yellowfin	*Limanda aspera*	
Snapper, Spotted Rose	*Lutjanus guttatus*			Sorubim, Barred	*Pseudoplatystoma fasciatum*	
Snapper, Star	*Lutjanus stellatus*			Sorubim, Spotted	*Pseudoplatystoma corruscans*	
Snapper, Timor	*Lutjanus timoriensis*			Sorubim, Tiger	*Pseudoplatystoma tigrinum*	
Snapper, Two-spot Red	*Lutjanus bohar*					

Spadefish, Atlantic	*Chaetodipterus faber*		
Spadefish, West African	*Chaetodipterus lippei*		
Spearfish, Longbill	*Tetrapturus pfluegeri*		
Spearfish, Mediterranean	*Tetrapturus belone*		
Spearfish, Roundscale	*Tetrapturus georgii*		
Spearfish, Shortbill	*Tetrapturus angustirostris*		
Spinefoot, Goldspotted	*Siganus punctatus*		
Spinefoot, Mottled	*Siganus fuscescens*		
Spinefoot, Orange-spotted	*Siganus guttatus*		
Spinefoot, Streaked	*Siganus javus*		
Spinefoot, Streamlined	*Siganus argenteus*		
Splake	*Salvelinus namaycush x S. fontinalis*		
Spot	*Leiostomus xanthurus*		
Spurdog, Japanese	*Squalus japonicus*		
Spurdog, Shortspine	*Squalus mitsukurii*		
Squirrelfish, Blue Lined	*Sargocentron tiere*		
Squirrelfish, Sabre	*Sargocentron spiniferum*		
Stargazer	*Uranoscopus scaber*		
Stargazer, Northern	*Astroscopus guttatus*		
Stargazer, Spotted Australian	*Ichthyscopus sannio*		
Steenbras, Red	*Petrus rupestris*		
Stingray, Atlantic	*Hypanus sabinus*		
Stingray, Black	*Dasyatis thetidis*		
Stingray, Bluespotted	*Neotrygon kuhlii*		
Stingray, Bluntnose	*Dasyatis say*		
Stingray, Common	*Dasyatis pastinaca*		
Stingray, Cowtail	*Pastinachus sephen*		
Stingray, Daisy	*Dasyatis margarita*		
Stingray, Diamond	*Dasyatis dipterura*		
Stingray, Estuary	*Dasyatis fluviorum*		

Stingray, Haller's Round	*Urobatis halleri*		
Stingray, Izu	*Hemitrygon izuensis*		
Stingray, Pelagic	*Pteroplatytrygon violacea*		
Stingray, Pitted	*Dasyatis matsubarai*		
Stingray, Red	*Dasyatis akajei*		
Stingray, Roughtail	*Dasyatis centroura*		
Stingray, Round	*Taeniura grabata*		
Stingray, S.A. Freshwater	*Potamotrygon motoro*		
Stingray, Sepia	*Urolophus aurantiacus*		
Stingray, Short-tail	*Dasyatis brevicaudata*		
Stingray, Southern	*Dasyatis americana*		
Stingray, Whip	*Dasyatis hastata*		
Stingray, W-mouth	*Dasyatis hypostigma*		
Sturgeon, Beluga	*Huso huso*		
Sturgeon, Lake	*Acipenser fulvescens*		
Sturgeon, Shortnose	*Acipenser brevirostrum*		
Sturgeon, Shovelnose	*Scaphirhynchus platorynchus*		
Sturgeon, White	*Acipenser transmontanus*		
Sucker, Blue	*Cycleptus elongatus*		
Sucker, Desert	*Catostomus clarkii*		
Sucker, Flannelmouth	*Catostomus latipinnis*		
Sucker, Klamath Largescale	*Catostomus snyderi*		
Sucker, Klamath Smallscale	*Catostomus rimiculus*		
Sucker, Largescale	*Catostomus macrocheilus*		
Sucker, Longnose	*Catostomus catostomus*		
Sucker, Lost River	*Deltistes luxatus*		
Sucker, Northern Hog	*Hypentelium nigricans*		
Sucker, Owens	*Catostomus fumeiventris*		
Sucker, Sacramento	*Catostomus occidentalis*		
Sucker, Sonora	*Catostomus insignis*		

Sucker, Spotted	*Minytrema melanops*		
Sucker, Utah	*Catostomus ardens*		
Sucker, White	*Catostomus commersonii*		
Sucker, Yaqui	*Catostomus bernardini*		
Sunfish, Green	*Lepomis cyanellus*		
Sunfish, Green X Redear	*Lepomis microlophus x cyanellus*		
Sunfish, Longear	*Lepomis megalotis*		
Sunfish, Redbreast	*Lepomis auritus*		
Sunfish, Redear	*Lepomis microlophus*		
Surfperch, Barred	*Amphistichus argenteus*		
Surfperch, Black	*Embiotoca jacksoni*		
Surfperch, Calico	*Amphistichus koelzi*		
Surfperch, Redtail	*Amphistichus rhodoterus*		
Surfperch, Rubberlip	*Rhacochilus toxotes*		
Surgeonfish, Black	*Acanthurus gahhm*		
Surgeonfish, Black-spot	*Acanthurus bariene*		
Surgeonfish, Eyestripe	*Acanthurus dussumieri*		
Surgeonfish, Indian Sail-fin	*Zebrasoma desjardinii*		
Surgeonfish, Orangeband	*Acanthurus olivaceus*		
Surgeonfish, Ringtail	*Acanthurus blochii*		
Surgeonfish, White-freckled	*Acanthurus maculiceps*		
Surgeonfish, Yellowfin	*Acanthurus xanthopterus*		
Swallowtail, Gorgeous	*Meganthias natalensis*		
Sweep	*Scorpis lineolata*		
Sweetlips, Painted	*Diagramma pictum*		
Sweetlips, Trout	*Plectorhinchus pictus*		
Swellshark, Australian	*Cephaloscyllium laticeps*		
Swordfish	*Xiphias gladius*		
Tai, Taiwan	*Argyrops bleekeri*		
Taimen	*Hucho taimen*		

Takenokomebaru	*Sebastes oblongus*		
Tambaqui	*Colossoma macropomum*		
Tanuki-Mebaru	*Sebastes zonatus*		
Tarakihi	*Nemadactylus macropterus*		
Tararira	*Hoplias lacerdae*		
Tarpon	*Megalops atlanticus*		
Tarpon, Oxeye	*Megalops cyprinoides*		
Tautog	*Tautoga onitis*		
Tench	*Tinca tinca*		
Tetra, Disk	*Myleus schomburgkii*		
Tetra, True Big-scale	*Brycinus macrolepidotus*		
Thornyhead, Shortspine	*Sebastolobus alascanus*		
Threadfin, East Asian	*Eleutheronema rhadinum*		
Threadfin, Giant African	*Polydactylus quadrifilis*		
Threadfin, Indian	*Alectis indica*		
Threadfin, King	*Polydactylus macrochir*		
Threadfin, Moi	*Polydactylus sexfilis*		
Threadfin, Smallscale	*Polydactylus oligodon*		
Threadfin, Striped	*Polydactylus plebeius*		
Threadfish, African	*Alectis alexandrina*		
Tigerfish	*Hydrocynus vittatus*		
Tigerfish, Campbells	*Datnioides campbelli*		
Tigerfish, Four-barred Siamese	*Datnioides polota*		
Tigerfish, Giant	*Hydrocynus goliath*		
Tigerfish, North African	*Hydrocynus brevis*		
Tilapia, Blue	*Oreochromis aureus*		
Tilapia, Hornet	*Tilapia buttikoferi*		
Tilapia, Mozambique	*Oreochromis mossambicus*		
Tilapia, Nile	*Oreochromis niloticus*		
Tilapia, Redbreast	*Tilapia rendalli*		
Tilapia, Spotted	*Tilapia mariae*		

Tilapia, Threespot	*Oreochromis andersonii*		
Tilefish, Atlantic Goldeye	*Caulolatilus chrysops*		
Tilefish, Bahama	*Caulolatilus williamsi*		
Tilefish, Blackline	*Caulolatilus cyanops*		
Tilefish, Blueline	*Caulolatilus microps*		
Tilefish, Golden-eyed	*Caulolatilus affinis*		
Tilefish, Great Northern	*Lopholatilus chamaeleonticeps*		
Tilefish, Red	*Branchiostegus japonicus*		
Tilefish, Saito's	*Branchiostegus saitoi*		
Tilefish, Sand	*Malacanthus plumieri*		
Tilefish, Southern	*Lopholatilus villarii*		
Toadfish, Gulf	*Opsanus beta*		
Toadfish, Leopard	*Opsanus pardus*		
Toadfish, Lusitanian	*Halobatrachus didactylus*		
Toadfish, Oyster	*Opsanus tau*		
Toadfish, Pacific	*Batrachoides pacifici*		
Torpedo, Atlantic	*Torpedo nobiliana*		
Torpedo, Spotted	*Torpedo marmorata*		
Trahira	*Hoplias malabaricus*		
Trahira, Giant	*Hoplias aimara*		
Treefish	*Sebastes serriceps*		
Trevally	*Pseudocaranx georgianus*		
Trevally, Australian Silver	*Pseudocaranx georgianus*		
Trevally, Bigeye	*Caranx sexfasciatus*		
Trevally, Blackbanded	*Seriolina nigrofasciata*		
Trevally, Blacktip	*Caranx heberi*		
Trevally, Blue	*Carangoides ferdau*		
Trevally, Bluefin	*Caranx melampygus*		
Trevally, Brassy	*Caranx papuensis*		
Trevally, Coastal	*Carangoides coeruleopinnatus*		

Trevally, Giant	*Caranx ignobilis*		
Trevally, Golden	*Gnathanodon speciosus*		
Trevally, Hampl's	*Caranx melampygus x sexfasciatus*		
Trevally, Imposter	*Carangoides talamparoides*		
Trevally, Island	*Carangoides orthogrammus*		
Trevally, Longfin	*Carangoides armatus*		
Trevally, Malabar	*Carangoides malabaricus*		
Trevally, Orange Spotted	*Carangoides bajad*		
Trevally, White	*Pseudocaranx dentex*		
Trevally, Whitefin	*Carangoides equula*		
Trevally, Yellowspotted	*Carangoides fulvoguttatus*		
Triggerfish, Blueline	*Xanthichthys caeruleolineatus*		
Triggerfish, Blunthead	*Pseudobalistes naufragium*		
Triggerfish, Bridled	*Sufflamen fraenatum*		
Triggerfish, Clown	*Balistoides conspicillum*		
Triggerfish, Finescale	*Balistes polylepis*		
Triggerfish, Grey	*Balistes capriscus*		
Triggerfish, Lagoon	*Rhinecanthus aculeatus*		
Triggerfish, Ocean	*Canthidermis sufflamen*		
Triggerfish, Orange-lined	*Balistapus undulatus*		
Triggerfish, Orangeside	*Sufflamen verres*		
Triggerfish, Pinktail	*Melichthys vidua*		
Triggerfish, Queen	*Balistes vetula*		
Triggerfish, Red-toothed	*Odonus niger*		
Triggerfish, Spotted Oceanic	*Canthidermis maculata*		
Triggerfish, Starry	*Abalistes stellaris*		
Triggerfish, Titan	*Balistoides viridescens*		
Triggerfish, White-banded	*Rhinecanthus aculeatus*		
Triggerfish, Yellowmargin	*Pseudobalistes flavimarginatus*		

Triggerfish, Yellowspotted	*Pseudobalistes fuscus*		
Tripletail	*Lobotes surinamensis*		
Tripletail, Pacific	*Lobotes pacificus*		
Tropical Two-wing Flyingfish	*Exocoetus volitans*		
Trout, Apache	*Oncorhynchus apache*		
Trout, Aurora	*Salvelinus fontinalis timagamiensis*		
Trout, Biwamasu	*Oncorhynchus rhodurus*		
Trout, Brook	*Salvelinus fontinalis*		
Trout, Brown	*Salmo trutta*		
Trout, Bull	*Salvelinus confluentus*		
Trout, Cutbow	*Oncorhynchus mykiss x O. clarki*		
Trout, Cutthroat	*Oncorhynchus clarkii*		
Trout, Gila	*Oncorhynchus gilae*		
Trout, Golden	*Oncorhynchus aguabonita*		
Trout, Lake	*Salvelinus namaycush*		
Trout, Masu	*Oncorhynchus masou*		
Trout, Ohrid	*Salmo letnica*		
Trout, Rainbow	*Oncorhynchus mykiss*		
Trout, Red Spotted Masu	*Oncorhynchus masou macrostomus*		
Trout, Tiger	*Salmo trutta x Salvelinus fontinalis*		
Trunkfish	*Lactophrys trigonus*		
Tsumagurokajika	*Gymnocanthus herzensteini*		
Tuna, Bigeye Atlantic	*Thunnus obesus*		
Tuna, Bigeye Pacific	*Thunnus obesus*		
Tuna, Blackfin	*Thunnus atlanticus*		
Tuna, Bluefin	*Thunnus thynnus*		
Tuna, Dogtooth	*Gymnosarda unicolor*		
Tuna, Longtail	*Thunnus tonggol*		

Tuna, Pacific Bluefin	*Thunnus orientalis*		
Tuna, Skipjack	*Katsuwonus pelamis*		
Tuna, Slender	*Allothunnus fallai*		
Tuna, Southern Bluefin	*Thunnus maccoyii*		
Tuna, Yellowfin	*Thunnus albacares*		
Tunny, Little	*Euthynnus alletteratus*		
Turbot	*Scophthalmus maximus*		
Turbot, Diamond	*Hypsopsetta guttulata*		
Turbot, Spottail Spiny	*Psettodes belcheri*		
Tusk	*Brosme brosme*		
Tuskfish, Azurio	*Choerodon azurio*		
Tuskfish, Blackspot	*Choerodon schoenleinii*		
Tuskfish, Robust	*Choerodon robustus*		
Ugui	*Tribolodon hakonensis*		
Ukkarikasago	*Sebastiscus tertius*		
Umazura Hagi	*Thamnaconus modestus*		
Umeiro	*Paracaesio xanthura*		
Unicornfish, Bignose	*Naso vlamingii*		
Unicornfish, Bluespine	*Naso unicornis*		
Unicornfish, Orangespine	*Naso lituratus*		
Unicornfish, Reticulate	*Naso reticulatus*		
Unicornfish, Sleek	*Naso hexacanthus*		
Unicornfish, Spotted	*Naso brevirostris*		
Unicornfish, Spotted	*Naso maculatus*		
Utsubo	*Gymnothorax kidako*		
Velvetchin, Short Barbeled	*Hapalogenys nigripinnis*		
Vimba	*Vimba vimba*		
Vundu	*Heterobranchus longifilis*		
Wahoo	*Acanthocybium solandri*		
Wallago	*Wallago attu*		
Wallago, Leerii	*Wallago leerii*		
Walleye	*Sander vitreus*		

Warmouth	*Lepomis gulosus*		
Weakfish	*Cynoscion regalis*		
Weakfish, Acoupa	*Cynoscion acoupa*		
Weakfish, Boccone	*Cynoscion praedatorius*		
Weakfish, Green	*Cynoscion virescens*		
Weakfish, Gulf	*Cynoscion othonopterus*		
Weakfish, Stolzmanns	*Cynoscion stolzmanni*		
Weakfish, Striped	*Cynoscion reticulatus*		
Weakfish, Stripped	*Cynoscion guatucupa*		
Wedgefish, African	*Rhynchobatus luebberti*		
Weever, Greater	*Trachinus draco*		
Wels	*Silurus glanis*		
Wenchman	*Pristipomoides aquilonaris*		
Whipray, Pink	*Himantura fai*		
Whitefish, Bonneville	*Prosopium spilonotus*		
Whitefish, Broad	*Coregonus nasus*		
Whitefish, Humpback	*Coregonus pidschian*		
Whitefish, Lake	*Coregonus clupeaformis*		
Whitefish, Mountain	*Prosopium williamsoni*		
Whitefish, Ocean	*Caulolatilus princeps*		
Whitefish, Round	*Prosopium cylindraceum*		
Whiting, Blue	*Micromesistius poutassou*		
Whiting, European	*Merlangius merlangus*		
Wolffish, Atlantic	*Anarhichas lupus*		
Wolffish, Bering	*Anarhichas orientalis*		
Wolffish, Northern	*Anarhichas denticulatus*		

Wolffish, Spotted	*Anarhichas minor*		
Wolf-Herring, Dorab	*Chirocentrus dorab*		
Wrasse Redbreasted	*Cheilinus fasciatus*		
Wrasse, Ballan	*Labrus bergylta*		
Wrasse, Blue-throated	*Notolabrus tetricus*		
Wrasse, Brazilian	*Halichoeres brasiliensis*		
Wrasse, Crimson-banded	*Notolabrus gymnogenis*		
Wrasse, Cuckoo	*Labrus mixtus*		
Wrasse, Girdled	*Notolabrus cinctus*		
Wrasse, Humphead Maori	*Cheilinus undulatus*		
Wrasse, Maori	*Ophthalmolepis lineolata*		
Wrasse, Peacock	*Iniistius pavo*		
Wrasse, Ringtail Maori	*Oxycheilinus unifasciatus*		
Wrasse, Rock	*Halichoeres semicinctus*		
Wrasse, Sandager's	*Coris sandeyeri*		
Wrasse, Scarlet	*Pseudolabrus miles*		
Wrasse, Surge	*Thalassoma purpureum*		
Wrasse, Yellow-brown	*Thalassoma lutescens*		
Wrasse, Yellow-saddled	*Notolabrus fucicola*		
Wreckfish	*Polyprion americanus*		
Yanaginomai	*Sebastes steindachneri*		
Yellowfish, Largemouth	*Labeobarbus kimberleyensis*		
Yellowtail, California	*Seriola lalandi*		
Yellowtail, Southern	*Seriola lalandi*		
Zander	*Sander lucioperca*		
Zander, Volga	*Sander volgensis*		